ENVIRONMENTAL SCIENCE, ENGINEERING AND TECHNOLOGY

RELATIVE HUMIDITY: SENSORS, MANAGEMENT AND ENVIRONMENTAL EFFECTS

ENVIRONMENTAL SCIENCE, ENGINEERING AND TECHNOLOGY

Additional books in this series can be found on Nova's website under the Series tab.

Additional E-books in this series can be found on Nova's website under the E-books tab.

ENVIRONMENTAL SCIENCE, ENGINEERING AND TECHNOLOGY

RELATIVE HUMIDITY: SENSORS, MANAGEMENT AND ENVIRONMENTAL EFFECTS

JOHN M. CASTILLO
EDITOR

Nova Science Publishers, Inc.
New York

For permission to use material from this book please contact us:
Telephone 631-231-7269; Fax 631-231-8175
Web Site: http://www.novapublishers.com

Additional color graphics may be available in the e-book version of this book.

LIBRARY OF CONGRESS CATALOGING-IN-PUBLICATION DATA
Relative humidity : sensors, management, and environmental effects / editor, John M. Castillo.
p. cm.
Includes index.
ISBN 978-1-61761-734-8 (hardcover)
1. Humidity. 2. Water vapor, Atmospheric. I. Castillo, John M.
QC915.R45 2010
551.57'1--dc22
2010036159

Published by Nova Science Publishers, Inc. † New York

CONTENTS

PREFACE

Relative humidity is a term used to describe the amount of water vapor that exists in a gaseous mixture of air and water vapor. The relative humidity of an air-water mixture is defined as the ratio of the partial pressure of water vapor in the mixture to the saturated vapor pressure of water at a prescribed temperature. This book presents current research in the study of relative humidity, including humidity-sensitive and non-ohmic tin dioxide ceramics; optics of airborne particles under elevated humidity conditions; a review of relative humidity management and its environmental effects; relative humidity effects in a spa; and relative humidity modeled by a functional network.

Chapter 1 - The objective of this chapter is to examine the characteristics of the low-level tropospheric temperature inversions to a height of 4 km above a suburban site at some of the Arabian Gulf Countries by using 10-year records of daily upper-air meteorological data. The temperature profile data included in this chapter was derived from the radiosonde data collected by the Meteorology Department at different stations in the Arabian Gulf Countries. The study addresses the statistical analysis of inversion parameters and investigates whether changes in these parameters (i.e., depth and strength) have occurred during the last decade. Both surface-based and elevated inversions are investigated. The study is important since it is the first investigation that addresses the statistical analysis of the thermal inversion layers in the Gulf region.

The chapter indicated monthly change in inversion frequency, height, depth, and strength. The frequency of the occurrence of all inversions seems to be slightly stable through out the period of the study. Since the frequency of the elevated inversions is the complement of the surface-based inversions, the highest frequency of surface-based inversions in some months corresponded to the time of minimum elevated inversions in the others. Overall, most of the inversions tended to be surface-based in all included stations. The surface-based and elevated inversions vary from winter to summer months depending on the changes of the atmospheric weather parameters in all stations of the study areas. The elevated inversions were generally lower and weaker than the surface-based inversions. The chapter also indicated annual variations in the thermal inversions associated with inversion depth, height, temperature and strength across both the surface-based and elevated inversions through out the 10-year period at some of the studied stations (Saudi Arabia Station No. 41024, Kuwait Station No. 40582, Abu Dhabi Station No. 41217, Salalah Station No. 41316).

Extensive studies should be highlighted for the comparison of the results of the present study areas with the observations of inversion layer in other Middle East countries and other regions in the world to explore the link between the thermal inversions, the phenomenon of larger-scale global warming, climate change, and the causes of cyclones and natural disasters.

Chapter 2 - Such as any other element, water also is evaporated. But since water plays a key role in our life on earth, its evaporation is very important in many aspects in different scientific fields. Water vapor is the main source of many atmospheric activities. It has an important role in the hydrologic cycle and water budget which is considered as the main solution of different water management problems. Evaporation has the core role in hydro-climatologic studies, watershed management, dams and water reservoirs designing. Evaporation is the most important parameter in irrigation programming and many other agricultural activities. It is important not only to prevent loosing water in drought but also to manage the consumption of water by plants, since it is the main source of loosing of water by plants through transpiration from stomata. Evaporation is most considered by agronomists in arid and semi-arid regions because limitation water availability is very restricted and evaporation is extremely high. It is also an important parameter in wet climates, when the need for investigation into its role in the water balance has been appreciated for a long time (Lemeur and Zhang, 1990).

For considering evaporation in any field of study we first need to know the basic physical theory of this phenomenon and also applying methods in different fields of studies.

Chapter 3 - Oxide ceramic semiconductors obtained by sintering of tin dioxide (SnO_2) with small additives of other oxides, exhibit strong deviations from Ohm´s law (varistor effect) and strong humidity dependence of conductivity at low electric fields (humidity sensor effect). SnO_2-based ceramics are suitable for the preparation of a device with double function (varistor and humidity sensor). Electrical conduction in such n-type semiconductors is controlled by the grain-boundary potential barriers formed as a result of sintering in oxidizing atmosphere. Humidity-sensitive and non-Ohmic properties of SnO_2 ceramics are explained by the decrease of the barrier height with the relative humidity and the electric field, respectively. Starting from these assumptions the functional dependences of electrical conductivity on relative humidity and on electric field are calculated and compared with experimental data. It is shown that electrical conductivity exponentially strongly depends on the relative humidity and the electric field.

Chapter 4 - The amount of radiative energy reaching the Earth's surface depends on many factors. Except for geometrical relations (such as a distance Earth-Sun, elevation of the Sun, and orography of the terrain surrounding a measuring site), the direct solar radiation as well as the diffuse radiation undergo complex alteration with regard to the actual meteorological conditions and the physical state of the atmosphere. Under clear sky conditions, the attenuation of the visible radiation is mainly due to scattering and absorption by air molecules and aerosol particles. These processes are linked, but, especially aerosols are responsible for rapid changes of the radiative fluxes. The effect of aerosols on the solar radiation becomes more important under elevated relative humidity (RH) conditions. The size of aerosol particles grows significantly when RH > 75% and the solid particles made up of the deliquescent materials may transform to solution droplets. Thus during the humidification process the optical properties of aerosol particles evolve. Definitely, the optical responses of humid particles will differ from those of non-humid particles. Therefore the RH is one of the crucial factors affecting the availability of radiative energy at the ground, implying a necessity to incorporate RH into optical models of the atmosphere. In this chapter the optical effects by single aerosol particles and particle polydispersions are discussed and some numerical demonstrations are shown.

Chapter 5 - Climate plays an important role in modulating reproductive success in cattle. Lactating dairy cows are under high metabolic demands due to high milk production and in this situation stressful factors such as climate conditions can compromise reproductive functions. Environmental temperature, radiant energy, relative humidity, wind speed and rainfall could contribute to the degree of ambient stress. To assess the influence of relative humidity (RH) and ambient temperature, the use of a temperature- humidity index (THI) which incorporates the effects of both is widespread. Several studies have related the THI index with fertility. In these studies, it is noteworthy that fertility decreased from 34% to 25% when THI value were higher than 75 on the day of insemination, or that THI values higher than 80 three days before insemination decrease fertility from 30-34% to 23%. On the other hand, prenatal loss is also an important factor affecting reproductive performance in high producing dairy cattle and high values of temperature-humidity index (THI) on days 21 to 30 of gestation have been found to be a risk factor for subsequent early foetal loss. Also, cumulative stressful climatic conditions and episodes of acute variation of climatic conditions in the first 40 days of gestation can compromise the success of gestation during both the cool and warm period of the year. A one-unit increase in the cumulative number of hours with THI > 85 during the days 11-20 of gestation period caused a 1.57-fold increase in the pregnancy loss, whereas the likelihood of foetal loss increased by a factor of 1.16 for each additional episode of acute increases in maximum THI values during gestation days 0-40. Finally, it is reported that during the second trimester of gestation, increased days with low RH in N. caninum-infected parous cows and heifers can compromise the success of gestation in a dry environment.

Chapter 6 - Three parameters relative humidity, temperature and organic substances are essential for the growth of molds. Building material like sandstone, marble, granite etc. are the inorganic material on which the growth of molds and other microbes is not favored, It was likely that organic residues from different sources on stone surface were utilized as nutrients. Light, temperature, humidity combined with rich nutrients of organic residues present on stone surface create suitable conditions for reproduction of dormant mold spores.

To study the effect of relative humidity – molds were isolated from three monuments Agra fort, Christian Cemetry and Kailash temple in three seasons (summer, rain and winter with relative humidity 50–60%, 80-98%, 70-85% respectively). The data showed maximum fungal diversity in rain season followed by winter season. In summer season, thermophilic fungi appeared in the procedure of isolation. Besides the deteriorated sandstone samples were also used for direct isolation of molds.

The effect of relative humidity was also confirmed by using tropical chamber test. This test used saturated chemicals to maintain definite relative humidity (like 96.6%, 85%, 76%). During the experimental study scanning electron microscopy was used for the analysis of pattern of biodeterioration. Many sporulating fungi and filamentous fungi were found to play active role in diagenesis of stone surface. A network of fungal hyphae penetrated the sandstone substrate and detached the loose arranged sand grains. The fungi bioweatherd silicates and other minerals through its biochemical and biomechanical activity. Besides, some crystals have also been found to be attached on their surface, which has been shown by SEM and XRD.

Chapter 7 - In atmospheric corrosion the time of wetness (TOW) refers to the time when the metal surface is sufficiently wet for corrosion to occur. It depends on the climatic conditions and on the nature of the surface at hand, and usually is expressed as a function of relative humidity (RH) and temperature (T). According to ISO 9223 standard, atmospheres with concentration of SO_2 lower than 10 mg/m^{2}day and concentration of chlorides lower than 3 mg/m^{2}day are considered as low polluted environments, and the effect of these atmospheric agents in the corrosive attack is considered as insignificant. In this type of atmospheres, corrosion rate of metals is closely related with this parameter, which is generally high in tropical atmospheres. However, research works about the relationship between TOW and corrosion of metals in tropical atmospheres with low concentrations of SO_2 and chlorides are very scarce.

In this study, carbon and galvanised steel were exposed for 14 months (between 2006 and 2007) at 9 different tropical sites in Colombia. These sites have low contents of SO_2 and chlorides, and have RH values higher than 60%. RH and T were recorded continuously in all sites, and TOW was calculated from these two parameters. Finally, models for the corrosion rate (mass loss) of both metals as a function of TOW and exposure time were obtained. Also, the applicability and limitations of the ISO 9223 standard in these atmospheres are discussed.

Chapter 8 - It is understood that indoor air must be classified and analysed based on different criteria like work risk prevention, energy saving, thermal comfort, biological sources, materials conservation and therapeutic effect. This study, in accordance with this classification, shows the indoor conditions that exist in different indoor ambiences, such as industrial environments, offices and school buildings, flats, libraries and a spa as well, using dataloggers that sample indoor conditions for long periods of time. Results revealed different limits of temperature and relative humidity, in concurrence with these criteria, and consequently, different procedures to control indoor ambiences. It is proposed, in particular, to define good behaviour habits in flats and libraries, like open windows for some time spans, to prevent the generation of excess humidity and improve the effect of passive methods like thermal inertia and permeable internal coverings in offices and schools. Hot ambiences especially, reveal two different control processes. On the one hand, in a spa, a low level of air change is proposed for energy saving, and low air velocity to prevent local thermal discomfort over wet skin surfaces. On the other hand, in industrial ambiences, it is proposed to increase the degree of air changes and enforce resting periods in a controlled ambience, to prevent work risk.

Chapter 9 - This chapter covers the indoor conditions of temperature and relative humidity in the special indoor ambience of a spa. The effect of these last two parameters, in particular, thermal comfort and energy saving will be analysed, revealing the main tendencies in accordance with the prescribed standards. Results showed that indoor air temperature depends on the pool water temperature, and that this in turn depends of the kind of health effect that we are looking at. Consequently, energy optimization is limited to spa application. However, the percentage of dissatisfied clients and indoor air acceptability showed a certain local discomfort in this ambience; therefore, a low residence time was required. Finally, some energy saving procedures were proposed to control the relative humidity of the ambience in the pool, like utilizing a protective covering over the pool, when not in use. Research works in future were proposed to control all these parameters based on the new systems to control temperature and relative humidity as well as new software resources.

Chapter 10 - The hygroscopic property (water uptake property at elevated relative humidity) of atmospheric aerosols is essential to understand the effect of aerosols on cloud formation by acting as cloud condensation nuclei (CCN) (i.e., hygroscopic particles grow into the CCN by absorbing water), which leads to climate change with cloud albedo effect. Also, hygroscopic property of particles is important to determine their transport behaviors and fates in the ambient atmosphere and to understnd their deposition pattern in the human respiratory system when they were inhaled. In this chapter, first, we describe a hygroscopicity tandem differential mobility analyzer (HTDMA) system in details to measure hygroscopic property of atmospheric aerosols in real time by measuring particle size change at an increased relative humidity. Then, we applied the HTDMA system to determine hygroscopic growth factors (i.e., ratio of particle size increase under high relative humidity) of various laboratory-generated particles of known composition and size to evaluate the performance of the HTDMA system. Finally, the hygroscopic GFs of particles in the ambient atmosphere at the relative humidity of 85% in various atmospheric environments (urban, rural, and coastal sampling sites) were measured and their results were compared. This will provide the estimation of chemical constituents of atmospheric aerosols in different environments, leading to a better understanding of their source characteristics.

Chapter 11 - The demand for high performance buildings, with regard to thermal comfort and energetic efficiency, is causing a progressive reduction of the envelope's air permeability, the increase of indoor RH% in occupied buildings and the appearance of problems linked with high humidity levels (deterioration of building materials, mould growth and biological organisms proliferation, respiratory discomfort, skin humidity and unacceptable perceived air quality).

This research investigates the possibility to use highly absorbing materials to dampen indoor RH% variations. The practical MBV of sodium polyacrilate, cellulose-based material, perlite and gypsum is evaluated for a daily cyclic exposure that alternates high (75%) and low (33%) RH% levels for 8h and 16h respectively. The adjustment velocity to RH% variations and the presence of hysteretic phenomena are also presented. The cellulose-based material proves to be the most suitable for moisture buffering applications. Starting from this material's properties, the effect of thickness, vapour resistance factor (μ) and mass surface exchange coefficient (Z_v) on sorption capacity is evaluated by the use of a numerical model.

Chapter 12 - The relative humidity will be modeled by a functional network. The relative humidity is based in real data obtained in a bird incubator. The functional networks compared with the neural networks have the advantage that they can find the global minimum while the neural networks can find a local minimum, but they have the disadvantage that they learn offline while some neural networks can learn online.

In: Relative Humidity: Sensors, Management… ISBN: 978-1-61761-734-8
Editor: John M. Castillo

Chapter 1

Analysis of Low-Level Temperature Inversions in Some of GCC Countries: Developing a Template for Characterization of Long Term Analysis of Climate Change and Global Warming in the Region

Sabah A. Abdul-Wahab* and Mohammed S. Al-Kalbani
Department of Mechanical & Industrial Engineering, College of Engineering, Sultan Qaboos University, Al Khoud, Oman

Abstract

The objective of this chapter is to examine the characteristics of the low-level tropospheric temperature inversions to a height of 4 km above a suburban site at some of the Arabian Gulf Countries by using 10-year records of daily upper-air meteorological data. The temperature profile data included in this chapter was derived from the radiosonde data collected by the Meteorology Department at different stations in the Arabian Gulf Countries. The study addresses the statistical analysis of inversion parameters and investigates whether changes in these parameters (i.e., depth and strength) have occurred during the last decade. Both surface-based and elevated inversions are investigated. The study is important since it is the first investigation that addresses the statistical analysis of the thermal inversion layers in the Gulf region.

The chapter indicated monthly change in inversion frequency, height, depth, and strength. The frequency of the occurrence of all inversions seems to be slightly stable through out the period of the study. Since the frequency of the elevated inversions is the complement of the surface-based inversions, the highest frequency of surface-based inversions in some months corresponded to the time of minimum elevated inversions in the others. Overall, most of the inversions tended to be surface-based in all included stations. The surface-based and elevated inversions vary from winter to summer months depending on the changes of the atmospheric weather parameters in all stations of the

* Corresponding author: Email: sabah1@squ.edu.om, Tel.: +968-2414-1360; fax: +968-2414-1316.

study areas. The elevated inversions were generally lower and weaker than the surface-based inversions. The chapter also indicated annual variations in the thermal inversions associated with inversion depth, height, temperature and strength across both the surface-based and elevated inversions through out the 10-year period at some of the studied stations (Saudi Arabia Station No. 41024, Kuwait Station No. 40582, Abu Dhabi Station No. 41217, Salalah Station No. 41316).

Extensive studies should be highlighted for the comparison of the results of the present study areas with the observations of inversion layer in other Middle East countries and other regions in the world to explore the link between the thermal inversions, the phenomenon of larger-scale global warming, climate change, and the causes of cyclones and natural disasters.

Keywords: Atmospheric parameters; balloon-borne sensor (radiosonde); meteorological stations; relative humidity

1. INTRODUCTION

Thermal inversion layer is defined as a layer of the atmosphere where the temperature increases with height. It usually refers to an increase in temperature with increasing altitude, which is a departure from the usual decrease of temperature with height (Kahl, 1990; Kahl et al., 1996; Phillpot and Zillman, 1970). According to Devasthale et al. (2010), the thermal inversion layer plays a crucial role in various processes by controlling the transfer of mass and moisture fluxes through the lower troposphere.

Thermal inversion layer occurs when a layer of warm air is settled or blanketed over a layer of denser cold air. The lapse rate becomes negative or inverted from the normal state, making it possible for temperature to increase with altitude. The net result of inversions from an air quality perspective is a very stable atmosphere characterized by little vertical mixing. This layer prevents the rise of cooling air and traps pollutants beneath it. It therefore has the propensity to cause air pollution (Abdul-Wahab et al., 1999).

There are two types of thermal inversions; radiation and subsidence. Radiation inversions usually occur during the night when the ground cools via the release of long-wave radiation. This type of inversion is most likely to occur during nights with clear skies and calm or light winds, when there is little horizontal or vertical mixing. Radiation inversions are also more common during winter months, when there are relatively longer periods when the amount of out-going long-wave radiation exceeds in-coming solar radiation. On the other hand, subsidence inversions are generally found to be associated with high-pressure systems or anticyclones. They are caused by downward vertical motions in the atmosphere that are associated with high-pressure areas (Abdul-Wahab et. al., 2004).

The study of atmospheric pollutants has received considerable attention all over the world for two main reasons: (i) they influence the climate directly by altering the earth's radiative budget and (ii) space-time variations of pollutants provide unique information on atmospheric conditions that influence environmental pollution and air quality (Devara et al., 1994). Thermal inversions, atmospheric stability, and mixed layer height are probably the most important meteorological conditions controlling atmospheric pollution dispersion (Turner, 1970; El Desouky and Abdul-Wahab, 1988). Moreover, studies and modeling of dispersion require knowledge of these parameters (Abdul-Wahab et al., 1999; Abdul-Wahab and Al-

Alawi, 2002). Remote sensing systems (acoustic sounding, Doppler acoustic sounders) or conventional in- situ systems (instrumented tower, tethered balloons, or instrumented aircraft) can be very useful to determine these variables (AeroVironment, Inc., 1976; Russell and Uthe, 1978; Gaffen et al., 1992a; 1992b).

Because of the importance of the inversion layer and its link with pollution, greenhouse gases and climate, it is important to study the behavior and the variations in its general characteristics. The increased concentrations of greenhouse gases have caused increases in global mean surface temperatures. Therefore, many investigators have studied the impact of climate change due to green house gases. Furthermore, surface temperatures affect the strength and the depth of the surface-based inversion layer. The presence of strong surface-based inversion layer, for example, is likely to cause energy to be trapped near the surface. This is an indication that changes in surface temperatures as well as climate change can be linked to the characteristics of the inversion layer (Abdul-Wahab et al., 2004).

For the past half century, radiosondes have been the main source of information on the state of the troposphere and lower stratosphere. About 2000 radiosondes are released every day from approximately 800 sites in the global aero-logical network to measure atmospheric temperature, pressure, humidity and winds. By virtue of global telecommunication networks and modern computer technology, radiosonde data from around the world are available to the international meteorological community. The accuracy of radiosonde measurements of temperature, pressure, relative humidity, and wind is of great importance to operational meteorologists. This is why International Departments of Meteorology carry out laboratory and flight tests using specialized reference equipment to determine radiosonde performance characteristics (Zaitseva, 1993).

Abdul-Wahab (2003) investigated the role of the elevated stable layer, particularly during the monsoon season in the Southern Region of the Sultanate of Oman, with particular interest in determining how the monsoon was dominated by the thermal inversions. Abdul-Wahab et al. (2004) also analyzed the inversion characteristics above a suburban site in the Sultanate of Oman by using 10-year records of data. Both surface-based and elevated inversion layers were investigated to examine whether the characteristics of these layers have changed during the past 10 years. The study reviewed the meteorological records with a view to setting up a database to look at the long-term trends in temperature due to climate change. In another study, Abdul-Wahab et al. (2005) investigated the relationships between the characteristics of thermal inversion and the health of people, especially the incidence of diseases related to atmospheric air pollution. Asthma, and other chronic obstructive lung diseases, heart failure, conduction and cardiac dysrhythmias, and ischemic heart diseases were among the diseases considered in their study. Emphasis was also placed on the temperature profile data to study the characteristics of inversions. Surface-based inversions occurred almost every day of the year while cases of elevated inversions were more during the summer than during the winter months. The study found that there were significant associations between the two types of inversions and the incidence of the diseases (Abdul-Wahab et al., 2005).

This chapter will address the low-level tropospheric temperature inversions above suburban sites in some of the Arabian Gulf Countries (Kuwait, Oman, Saudi Arabia, and United Arab Emirates) using daily upper-air meteorological data. The work is important since it is the first investigation that addresses the statistical analysis of the thermal inversion layers in the Gulf region, which is projected to compound the pressures on natural resources and the environment, associated with rapid urbanization, industrialization and economic development. The aims of the chapter are to study the characteristics of long-term analysis of climate change and global warming through the determination of low-level temperature inversions, using radiosonde measurements in the Arabian Gulf Countries.

2. Previous Literature Preview

Many studies have dwelt on the physical controls behind the inversion layer (Busch et al., 1982; Overland and Guest, 1991; Overland and Davidson, 1992). Chang et al. (1980), for example, stressed the role of thermal inversion in explaining the monthly variation of NO_2 and O_3. Another factor for explaining the seasonal variability was the depth of the mean surface-based inversion and inversion layer stability (Haagenson, 1979). It has also been established that a combination of higher stability of surface-based inversion and less surface heating from sunshine to the morning rush hour during the colder months, resulting in a shallow mixing layer, was responsible for the higher pollutant levels in the winter months (Chang et al., 1980). Kelly et al. (1982) described annual and seasonal changes in air temperatures over high latitudes of the Northern Hemisphere during the period 1881-1980. Trends in the average temperature of the Arctic were compared with those of the Northern Hemisphere. Seasonal and regional departures from the long-term trends in the average temperature of the Arctic were identified. Spatial patterns of variation in the Arctic temperature field were determined by principal component analysis and the major characteristics of the time series of the dominant patterns were summarized. The Arctic variations were greater in magnitude and more rapid. The results also showed that the trends in general have been Arctic-wide, but that certain regions were particularly sensitive to long-term variations.

In one of the early papers on inversion climatology using a 2-year period of radiosonde data, Hosler (1961) acknowledged the effect of the anticyclonic cell off the coast of California on the climatology of inversions in California and the southern part of Oregon. Bilello (1966) investigated inversion characteristics such as frequency, base height, thickness, base temperature, and temperature gradient. Using a 4-year period of radiosonde data, Tyson et al. (1976) tried to establish the climatology of surface inversions over Southern Africa. They recognized that the movement of high and low pressure systems as well as the south Atlantic anticyclone were the major factors affecting the spatial and temporal variations in the characteristics of surface inversions. Similarly, for elevated inversions over southern Africa using the same 4-year period data as in Tyson et al. (1976), Preston-Whyte et al. (1976) found that the eastward moving high and low pressure systems, and the south Atlantic anticyclone are again the primary controls affecting the spatial and temporal variations of the characteristics of elevated inversions.

A network of 63 well-distributed radiosonde stations was used to estimate mean- annual temperature variations at the surface and for 85-30 kPa (850-300 mb), 30-10 kPa, 10-5 kPa, 10-3 kPa, and surface 10 kPa layers for five climatic zones, both hemispheres, and the world for the interval 1958-1981(Angell and Korshover, 1983). Angell (1988) examined the variations and trends in tropospheric and low-stratospheric temperature for seven climatic zones, hemispheres, and world for intervals 1958-1987 and 1973-1987, based on 63 well-distributed radiosonde stations. Karoly (1989) used the radiosonde temperature data from 147 stations in the Northern Hemisphere for the period 1964-1985 to investigate recent temperature trends in the troposphere and lower stratosphere. Experiments with atmospheric general circulation models indicated that increased concentrations of greenhouse gases in the atmosphere will lead to reduced temperatures in the stratosphere as well as increased temperature in the troposphere. An index designed to identify this signal was computed from the station data. This index was found to have a positive trend at most of the stations, consistent with the greenhouse effect. Over the hemisphere, the signal was statistically significant.

Timerev and Egorova (1991) studied the spatial-temporal variability of surface inversions in the Arctic using aero-logical soundings form drifting (1954-1988), and stationary Soviet and foreign stations (1961-1970). They determined the frequency and thickness of surface inversions to specify areas with the largest surface aerosol concentration. They also reported that the inversion layer over the frozen Arctic Ocean has become shallower and more thermally stable.

Climatological characteristics of the low-level tropospheric temperature inversion in the Canadian Arctic were examined using 10-40 records of upper-air meteorological data (Kahl et al., 1992). Inversions at the northern sites were primarily surface-based in winter and elevated from mid-spring through summer. At the southern sites, a bimodal pattern was observed with surface-based inversions occurring during late summer, as well as during winter.

Serreze et al. (1992) introduced the idea of the complexity of inversions, suggesting that inversions were not only affected by radiative cooling, but also by warm air advection, subsidence, radiative properties of ice crystals, surface melt, and topography. They studied the Arctic low-level temperature inversion using up to 12 years of twice-daily radiosonde data from 31 inland and coastal sites of the Eurasian Arctic and a total of nearly six years of data from three Soviet drifting stations near the North Pole. The authors found that the frequency of inversions, the median inversion depth, and the temperature difference across the inversion layer increased from the Norwegian Sea eastward toward the Laptev and East Siberian seas. The inversions tended to be most intense during winter due to the large deficit in surface net radiation. The frequency of inversions was lowest during summer and most inversions were elevated. In their study, they showed that the parts of Eurasian Arctic Ocean experience inversion frequencies greater than 80% even in summer, while the frequencies were very close to 100% in winter (especially over inner Arctic).

The Arctic region has come to be viewed as a sensitive indicator of global climate change. Kahl et al. (1993) examined Arctic temperature trends in four tropospheric layers during the period 1958-1986 through the analysis of a comprehensive archive of Arctic upper air meteorological data. On the basis of their analysis, they concluded that greenhouse-induced warming was not detectable in the Arctic troposphere for the 1958-1986 periods. In another relevant study on the Arctic, the variations in surface air temperatures in the Arctic during the period 1979-1997 were analyzed by Rigor et al. (2000). The study concluded that the changes in surface air temperature over the Arctic Ocean were related to the Arctic oscillation, which accounts for more than half the surface air temperature trends over Alaska, Eurasia, and the eastern Arctic Ocean but less than half in the Western Arctic Ocean.

Milionis and Davies (1992) studied the temperature inversions up to the 500-mbar level in 50-mbar depth layers over Hemsby, in eastern England. The inversions were characterized by vertical gradients of temperature, potential temperature, dew point temperature, and depth. An inversion index, which is the product of potential temperature gradient and frequency of occurrence, was used. They attributed the maximum of the vertical activity of inversions, which was found to be in a layer from about 950-800 mbar, to anticyclonic subsidence. Inversion characteristics were essentially similar for day and night, except in the 1000-950 mbar layer, where daytime frequency was greater.

Gaffen et al. (1992a) quantified the relationships between surface air temperature and precipitation of water vapour for different time scales using radiosonde data from 50 stations from 1973 to 1990. They showed that the relationship between the precipitation water vapour and surface air temperature was influenced by the time scales and geographic region. Bradley et al. (1992) investigated the climatology of surface-based inversions in the North American Arctic. They found that the inversion frequency was at a maximum in winter when inversions range from 400 to 850 m thickness. Inversion thickness and strength were strongly related to surface temperature. Midwinter inversions commonly persisted for 2-4 days, but may remain undisturbed for several weeks, affecting lower tropospheric chemistry.

The boundary layer structure measured by 402 pairs of approximately collocated radiosonde and dropsonde temperature profiles over the Arctic Ocean during the period 1957-1961 was examined by Skony et al. (1994). The results indicated that although they contained relatively few measurement levels, the dropsonde soundings contain sufficient vertical resolution to characterize the temperature inversion. Gaffen, (1994) combined historical information on changes in radiosonde instruments and observing method with time series of upper-air temperature data to estimate the effects of changes in sensors; changes in solar radiation corrections to the data; and changes in the length of the train between the balloon and the instrument package. These changes can induce discontinuities in the temperature records from several tenths to as high as several degrees Celsius.

Long-term variability in the low-level inversion layer over the Arctic Ocean was investigated by Kahl et al. (1996) using nearly 30 000 Arctic Ocean temperature profiles for 1950-1990. The authors concluded that the Arctic inversion layer had changed significantly during the past four decades, becoming stronger, occurring more frequently, and more frequently becoming elevated above the surface. Zaitseva et al. (1996) investigated temperature inversions over the western arctic from radiosonde data.They used a unique bank of radiosonde data on air temperature (from surface to 3-5 km) over the western Arctic Ocean to analyze a thermal structure of the lower troposphere. A way of this, Sultan and Elghribi (1996) studied the temperature inversion in the Arabian Gulf and the Gulf of Oman

The radiosonde temperature measurements in strong inversions were carried out by Mahesh et al. (1997). The authors analyzed the radiosonde temperature measurements at the South Pole, where steep surface-based temperature inversions occur during most of the year. They focused on the quantification and correction of thermal lag of the thermistor. To this end, they developed a procedure to correct the temperature data from routine radiosonde soundings for the finite response time of the thermistor. The results showed that the equilibrated temperatures measured by the tethered sonde at a given pressure level were higher than those from the balloon-borne sound throughout most of the inversion layer. They recommended that a correction for thermal lag should be done for all situations where radiosondes were used to measure steep temperature gradients in the boundary layer.

The use of radiosonde temperature data in climate studies was studied by Luers and Eskridge (1998). For this effect, ten of the most common radiosondes used throughout the world since 1960 have been evaluated concerning potential use of their temperature data for climate studies. Those ten radiosondes were evaluated by modeling the temperature of the sensing element relative to the temperature of the air in which it is immersed. Validation and sensitivity analysis studies were performed on each radiosonde model as a means of estimating the environmental parameters that influence the temperature error and the resulting accuracy of the day and nighttime temperature profiles.

The formation and destruction of inversion layers within a deep valley was addressed by Anquetin et al. (1998). The study highlighted the influence of the season on the mechanisms responsible for the formation and the destruction of the thermal inversion layer. Moreover, Galerkin et al. (1998) highlighted the climatic characteristics of the layer of thermal inversions in the ocean. In their own research in the field, Majorowicz and Safanda (1998) studied the ground surface temperature from inversions of underground temperature in the Western Canadian Sedimentary Basin. Using radiosonde data for a 15-year period, Prezerakos (1998) tried to associate the characteristics of temperature inversions over Athens with the synoptic-scale conditions in a qualitative manner. Also, Whiteman et al. (1999) related the characteristics of temperature inversions with the prevailing weather conditions. They argued that the evolution of temperature inversions during winter in the Colorado plateau basin is driven by synoptic scale events.

The associations between atmospheric temperature inversions and vertical wind profiles was assessed by Milionis and Davies (2002). A preliminary statistical assessment was made, using message data from radiosonde soundings at a station in eastern England for the period 1976-1980, from the surface up to 700 mb. Detection and analysis of clear-sky, low-level atmospheric temperature inversions with Moderate Resolution Imaging Spectroradiometer (MODIS) were studied by Liu and Key (2003). They developed a method for detecting and estimating the characteristics of clear-sky, low-level temperature inversions using the MODIS on the Terra and Aqua satellites.

The temperature, humidity, and pressure response of radiosondes at low temperatures were investigated by Hudson et al. (2004). The response of radiosondes to an instantaneous change of environment was studied by taking the instruments from a warm building into the cold environment at South Pole Station. After being initialized inside, the radiosondes were carried outside and placed on the snow surface, where they were left until they reported stable values of temperature, pressure, and relative humidity. Hudson et al. (2004) studied the response of radiosondes change of environment by taking the instruments from a warm building into the cold environment at South Pole station. The study concluded that either storing and preparing radiosondes in ambient conditions, or allowing them at least 30 minutes to equilibrate after being taken outside from a warm building, could make a substantial improvement in the quality of upper-air data collected at cold stations. Intercomparison of atmospheric soundings from the aerosonde and radiosonde was presented by Soddell et al. (2004). Although the Aerosonde generally performed well in comparison with the radiosonde, calibration errors and time-lag errors similar to those observed between radiosonde and dropsonde observations were evident in some of the temperature and relative humidity profiles. Chen et al. (2004) conducted a study to compare the Rayleigh lidar system with radiosonde measurements. Good harmony between these two techniques was found. The study indicated that lidar technique was more accurate in locating the height and temperature of the tropopouse when compared to radiosonde observations.

Hudson and Brandt (2005) looked at the surface-based temperature inversion on the Antarctic Plateau. Data from radiosondes, towers, and a thermistor string were used to characterize the temperature inversion at two stations. Ten years of temperature data from a

22-m tower at the South Pole were analyzed. The results showed that the minimum temperature and the maximum inversions tended to occur not with calm winds, but with winds of 3-5 ms^{-1}, due likely to the inversion wind. The monthly mean temperature profiles showed large variations in lower stratospheric temperatures, and in the strength and depth of the surface-based inversion. The effects of the diurnal cycle of solar elevation over the Antarctic Plateau showed that strong inversions develop quickly at night when the sun is low in the sky, but are destroyed during the middle of the day.

Gopalakrishna et al. (2005) investigated the observed variability of sea surface salinity and thermal inversions in the Lakshadweep Sea during contrast monsoons. Kassomenos and Koletsis (2005) studied the seasonal variation of the temperature inversions over Athens (Greece). They used the methodology developed by Milionis and Davies (1992) to examine the seasonal variation of temperature inversions over Athens using a 25-year period of radiosonde data. The study examined the frequency of appearance and the characteristics of temperature inversions over Athens, up to the pressure level of 500 hPa, for the time period 1974-2001 on a seasonal basis. They attempted in a qualitative way to relate the formation of inversions in each season with the synoptic conditions. In their study of air pollution episodes in four European cities, Kukkonen et al. (2005) associated the presence of persisting inversions with synoptic systems of high pressure and high-pressure ridges.

Validation of atmospheric Infrared Sounder temperature and water vapor retrievals with matched radiosonde measurements and forecasts was evaluated by Divakarla et al. (2006). Kurian and Vinayachandran (2006) studied the formation mechanisms of temperature inversions in the southeastern Arabian Sea (SEAS) during winter. For this purpose, numerical simulations using a high resolution (0.25° × 0.25°, 40 levels) OGCM of the Indian Ocean were used. Janhäll et al. (2006) studied the evolution of the urban aerosol during winter temperature inversion episodes. Using radiosonde data for a period of 29 years from upper air stations in Indochina Peninsula, Nodzu et al. (2006) studied the interaction between temperature inversions and seasonal changes in convective activity during the dry season to rainy season transition. In a study of a severe air pollution episode in Logan at Utah in USA, Malek et al. (2006) associated the presence of a strong temperature inversion with a high-pressure weather system.

Haimberger (2007) introduced a new technique that uses time series of temperature differences between the original radiosonde observations (obs) and background forecasts (bg) of an atmospheric climate data assimilation system for homogenization. The systematic differences in aircraft and radiosonde temperatures were presented by Ballish and Kumar (2008). The paper compared temperatures of aircraft and radiosondes by collocation and mean temperature observation increments differences (MTOI), along with discussing the pros and cons of each method, with neither providing an absolute truth. El-Askary and Kafatos (2008) investigated the thermal inversions as a major contributor to the black cloud episodes over Cairo. Zwoździak et al. (2008) investigated the morning ozone buildup during summer 2003 episodes and their links with some radiosonde data in WrocŁaw (South-Western Poland). A comparison of temperature inversion statistics at a coastal and a non-coastal location influenced by the same synoptic regime was examined by Milionis and Davies (2008a). The main objective of the work was to determine to what extent the climatology of atmospheric temperature inversions at one location is site specific, and to what extent it reflects a wider area for which the same synoptic conditions can be assumed. In another study, Milionis and Davies (2008b) expressed quantitatively the effect of the prevailing weather on the statistics of atmospheric temperature inversions. This effect stems from the fact that certain conditions that enhance, entail, or inhibit the formation of certain types of inversions, are linked with particular weather patterns. The results of their work can be used for the creation of correction factors for many characteristics related to inversion climatology. Such corrections are necessary in cases where the weather types in the study period do not reflect their frequency in the average climate.

A new look at radiosonde data prior to 1958 was carried out by Grant et al. (2009). In their study, the authors proposed corrections and presented corrected hemispheric fields that showed large changes and improved internal consistency in height and temperature across Eurasia compared with uncorrected data. Those corrections were important, especially as they had a clear spatial structure that interfered with the planetary wave structure. Pérez et al. (2009) classified CO_2 concentrations based on a binary meteorological six variable system. In their paper, a close relationship between CO_2 concentrations and meteorological variables was proposed to establish atmospheric conditions linked to high CO_2 values. Nisha et al. (2009) examined the reduced near-surface thermal inversions in 2005-06 in the southeastern Arabian Sea (Lakshadweep Sea). The study described the year-to-year variability in the observed near-surface thermal inversions during the winter seasons of 2002-2006. Chutko and Lamoureux (2009) examined the influence of low-level thermal inversions on estimated melt-season characteristics in the central Canadian Arctic. Surface and upper-air temperature data from Resolute, Cornwallis Island, was used to estimate vertical lapse rates up to 300 m asl, and to identify the presence of inversions at that altitude. Aikawa and Hiraki (2009) studied the characteristic seasonal variation of vertical air temperature profile in urban areas of Japan. Olofson et al. (2009) extended the previous study carried out by Janhäll et al. (2006). The aim of their study was to better understand the effects of boundary layer dynamics on the urban aerosol properties. They studied the influence of inversion episodes on urban aerosol properties and on the formation of new particles during wintertime temperature inversions in Gothenburg (Sweden). In northern Arizona of the United States, the nocturnal temperature inversions in a small, enclosed basin and their relationship to ambient atmospheric conditions were investigated by Yao and Zhong (2009).

Alaska surface-based temperature inversions from a climate perspective were analyzed by Bourne et al. (2010) using radiosonde observations from different climate zones in Alaska. Inversion climatology, variability and links to the large-scale climate were investigated for the period of 1957-2008 when high quality radiosonde data were available. Devasthale et al. (2010) quantified the characteristics of inversions in terms of frequency and strength for the entire Arctic Ocean for summer and winter seasons using the AIRS data for the clear-sky conditions.

3. METHODOLOGY

3.1. Selection of Observing Stations

This chapter was based on the study of statistical analysis of the upper air data collected from Kuwait, Oman, Saudi Arabia, and United Arab Emirates (Figure 1) over a period of 10 years starting in 1994 and ending in 2003. This period was selected because some observing stations in the Gulf Countries have only 10 years record. Therefore, to be consistence with other stations the chapter was concentrated on this study period. The statistical analysis was focused on temperature profile observations. The upper data was recorded by selecting airport observing stations from the Gulf Countries shown in Table 1. For Kingdom of Saudi Arabia, 3 observing stations (40416) located at 26 16 °N, 50 10 °E, (40417) located at 26 27 °N, 49 49 °E, and (41024) located at 21.42 °N, 39.11 °E were used. In the State of Kuwait, the observing station number (40582) located at 29.22 °N, 47.98 °E, was used. For United Arab Emirates, the data was recorded by the Abu Dhabi International Airport observing station (40582) which was located at 29.22 °N, 47.98 °E. In the Sultanate of Oman, the data was recorded by Salalah Airport observing station number (41316) located at 17.02 °N, 54.05 °E (Figure 2).

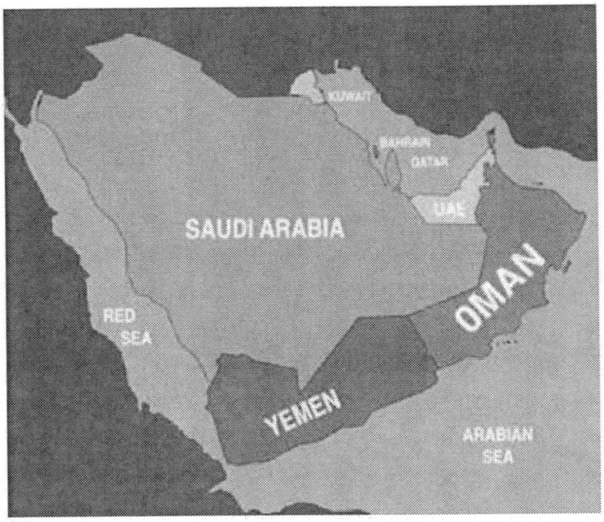

Figure 1. Map of the Arabian Peninsula

Table 1. World meteorological organization observing stations at Gulf countries

Station index number	Name and location	Country	Latitude	Longitude	Elevation (m)
40416	Dhahran	Kingdom of Saudi Arabia	26 16N	50 10E	17
40417	King Fahad International Airport Dammam	Kingdom of Saudi Arabia	26 27N	49 49E	12
41024	King Abdul Aziz International Airport Jeddah	Kingdom of Saudi Arabia	21 42N	39 11E	17
40582	Kuwait International Airport	State of Kuwait	29 22N	47 98E	55
41217	Abu Dhabi International Airport	United Arab Emirates	24 43N	54 65E	27
41316	Salalah Airport	Sultanate of Oman	17 02N	54 05E	22.84

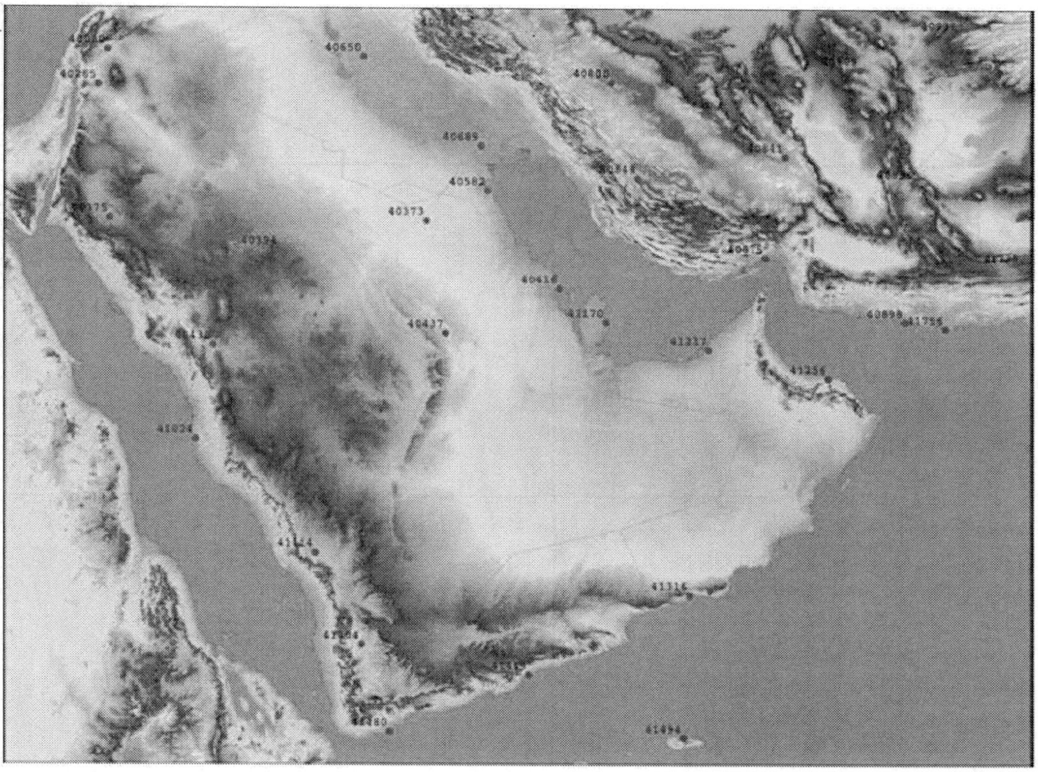

Figure 2. Map of world meteorological organization observing stations index. Numbers showing the stations of the study areas at different gulf countries

3.2. Experimental Design

The data were collected from a balloon-borne sensor (radiosonde) launched daily by the Meteorological Departments in different countries at 0230 a.m. local time (2230 UTC). A radiosonde is a small, lightweight package of instruments, suspended from a balloon. When the balloon was released, the apparatus was launched through the atmosphere at approximately 5 m s^{-1}. Measurements of pressure, temperature, humidity, wind speed and direction were taken at intervals during flight. Information was transmitted to the surface as a radio signal. The ground station equipment received the signal and processed the data. The data consisted of thermodynamic information from the surface to pressure of approximately 20 hpa. The flight of the balloon varied according to the strength of the wind and thus may not ascend strictly vertically. Therefore, the balloon will not always report information directly over the selected observing stations in the Gulf Countries. Nevertheless, when analyzing weather systems on the synoptic scale, the profiles were assumed to be vertical without significant error.

Since the processes within the mixing layer 1-2 km above ground level generally control ground-level pollutant concentrations and transport, the investigation of this study was limited mainly to the lower 4 km of the atmosphere. Therefore, the upper air data code format taken from the Meteorological department at the observing stations, were interpreted to have occurred at a height of about 4 km above the ground level.

3.3. Statistical Analysis of Data Collection

Emphasis was placed on the characteristics of temperature inversions. Therefore, the temperature profile data were derived and tabulated on a monthly and yearly basis. The statistical analysis of the temperature data was performed to determine the type, height and strength of the thermal inversions in the period covered by the study. The analysis of the thermal inversion data collected over previous years was then used indirectly to determine changes in the earth surface temperature and thus assessed climate variability as well. Statistical modelling was then applied to predict potential future changes in temperature variability of the earth surface. The statistical analysis of the data was performed using data analysis software SPSS (Statistical Package for the Social Sciences, Version 13). An important part of the study was to investigate the relationship between the occurrence of thermal inversions and atmospheric parameters by studying inversion frequencies, height, depth, and strength.

3.4. Determination of Inversion Parameters

For each studied station, inversions were identified using the detection algorithm developed by Kahl (1990). Each sounding was scanned upward from the surface, identifying the inversion base as the bottom of the first layer in which the temperature increases with altitude. The inversion top was defined as the bottom of the first adjoining layer in which the temperature decreased with altitude (Kahl et al., 1996). However, the thin lapse layers embedded within the overall inversion layer (i.e., positive lapse), which were calculated by Kahl (1990) were not allowed in this project. The technique for not allowing embedded lapse layers was explained by Bradley et al. (1993). Kahl et al. (1996) stated that considering these thin layers to be part of the overall layer would tend to increase the total temperature difference across the inversion.

For each sounding that contained an inversion, three parameters were computed: height of inversion base (altitude above ground level, Z_b), depth of the inversion layer (difference in altitude between the top and the base of the inversion, ΔZ), and temperature difference between the top and bottom of the inversion layer (temperature difference across the inversion layer, ΔT). The temperature difference was selected to characterize the strength (i.e., thermal stability) of the inversion.

To determine the frequencies with which surface-based and elevated inversions occurred each month in different stations, the equations described by Kahl et al. (1996) were used. The inversion frequencies were defined by Kahl et al. (1996) as follows:

$$N_{inv} = N_{SBI} + N_{elev} \tag{1}$$

$$F_{elev} = N_{elev} / N_{inv} \tag{2}$$

$$F_{SBI} = N_{SBI} / N_{inv} \quad (3)$$

$$F_{all} = N_{inv} / N_{sound} \quad (4)$$

Where:

N_{SBI} is number of surface-based inversions,
N_{elev} is number of elevated inversions,
N_{inv} is total number of inversions,
N_{sound} is the total number of soundings including those soundings in which an inversion was not detected,
F_{SBI} is frequency of occurrence of surface-based inversions,
F_{elev} is frequency of occurrence of elevated inversions, and
F_{all} is frequency of occurrence of all inversions.

4. RESULTS

4.1. Analysis of Thermal Inversions in Saudi Arabia (Station NO. 40416)

4.1.1. Inversion frequencies

The results of the calculations of the inversion frequencies, which occurred in the Kingdom of Saudi Arabia (Station NO. 40416) are shown in Table 2.

Table 2. Monthly inversion frequencies (Saudi Arabia Station Number 40416) during 1994 - 2003

Month	N_{SBI}	No surface-based inversion	N_{elev}	No elevated inversion	N_{inv}	N_{sound}	F_{SBI}	F_{elev}	F_{all}
1	106	16	23	99	129	244	82.2	17.8	52.9
2	68	12	20	59	88	159	77.3	22.7	55.3
3	99	22	34	87	133	242	74.4	25.6	55.0
4	100	15	57	58	157	230	63.7	36.3	68.3
5	111	4	75	40	186	230	59.7	40.3	80.9
6	80	39	11	108	91	238	87.9	12.1	38.2
7	108	15	1	122	109	246	99.1	0.9	44.3
8	110	13	0	123	110	246	100.0	0.0	44.7
9	110	9	46	73	156	238	70.5	29.5	65.5
10	82	10	46	46	128	184	64.1	35.9	69.6
11	75	12	15	72	90	174	83.3	16.7	51.7
12	83	10	21	72	104	186	79.8	20.2	55.9

$\mathbf{N_{SBI}}$ is number of surface-based inversions, $\mathbf{N_{elev}}$ is number of elevated inversions, $\mathbf{N_{inv}}$ is total number of inversions, $\mathbf{N_{sound}}$ is number of soundings, $\mathbf{F_{SBI}}$ is frequency of occurrence of surface-based inversions, $\mathbf{F_{elev}}$ is frequency of occurrence of elevated inversions, and $\mathbf{F_{all}}$ is frequency of occurrence of all inversions.

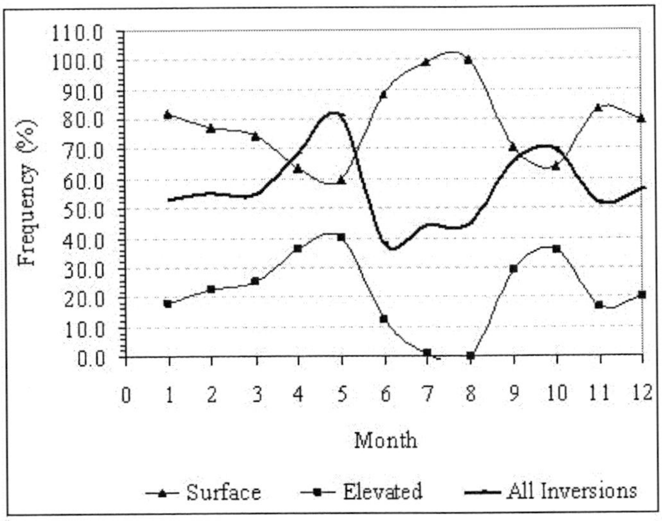

Figure 3. Monthly variations in inversion frequencies, given for each month the frequency of occurrence of surface-based inversions, the frequency of elevated inversions, and the frequency of all inversions **(Saudi Arabia Station Number 40416)**

Figure 3 shows the frequency of occurrence of surface-based inversions, the frequency of elevated inversions, and the frequency of all inversions. Figure 3 reveals that an inversion occurs in every month examined, except the month of August, which shows no occurrence of elevated inversion. Overall, most of the inversions tended to be surface based. On examining the curve of the frequency of all inversions (F_{all}), it seemed that the inversions were least frequent in summer months (June, July and August) while there appeared to be an increase in the frequency of (F_{all}) in the remaining months. The maximum frequency of 80.9% occurred during May, and minimum frequency of 38.2% occurred during June. Also, June, July and August showed the highest frequencies of surface based inversions compared to other months.

Since F_{elev} is the complement of F_{SBI}, an opposite trend in elevated inversions was observed. In addition, it can be seen that the highest frequency of surface-based inversions in some months corresponded to the time of minimum elevated inversions in the others.

4.1.2. Inversion height

The mean height of the inversion base and top according to the month of the year during the period and area of the study in Station Number 40416 is shown in Figure 4. Both surface-based and elevated inversions are included in this figure. The base of the surface-based inversion is defined by station 40416 elevation with respect to the mean sea level. It can be clearly seen that the heights of the top of the surface-based inversion layer reached the maximum depth during June, July and August, and almost the same as surface base during the remaining months of the year. On the other hand, the elevated inversions were frequent in all months and their heights (top and base) were at their maximum during the month of July. By comparing the heights of the inversion top and base of both the surface-based and elevated inversions, it can be seen that the month of July indicated the highest inversion.

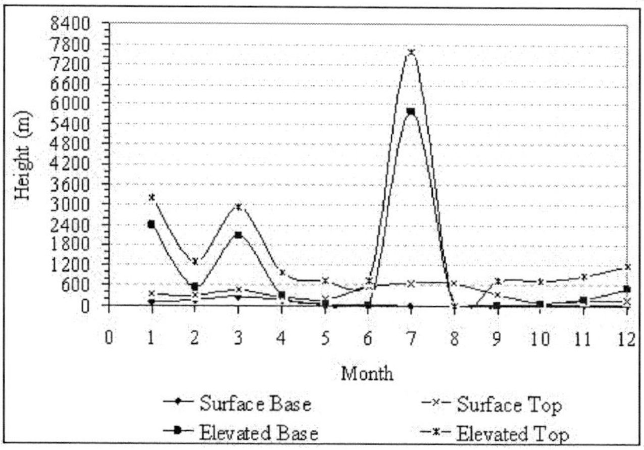

Figure 4. Mean height of the inversion base and top (m) according to the month of the year during the period of the study. Both surface-based and elevated inversions are included (**Saudi Arabia Station Number 40416**)

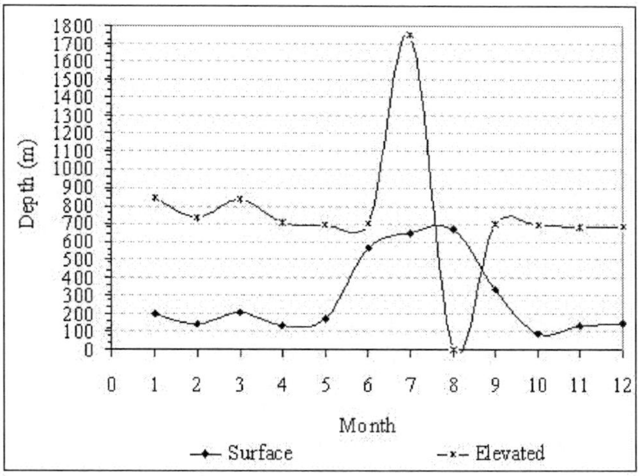

Figure 5. Mean inversion depth (m) according to the month of the year during the period of the study. Both surface-based and elevated inversions are included (**Saudi Arabia Station Number 40416**)

4.1.3. Inversion depth

The mean inversion depth was determined by calculating the difference in altitude between the top and the base of the inversion. The results of the mean inversion depth for both the surface-based and elevated inversions according to the month of the year during the period of the study in Station number 40416 are shown in Figure 5. The most obvious feature of this figure was that the elevated inversions were generally deeper than the surface-based inversion layers. This was seen in all months of the year, except during August where the mean depths of elevated inversion was zero. For the surface-based inversion, the mean depth inversion was maximum in months June, July and August, and the least in the remaining months with an average of about 100 - 200 m. For the elevated inversion, the deepest inversions occurred in July with an average value of 1750 m, and the least in the other months with an average values between 700 and 850 m.

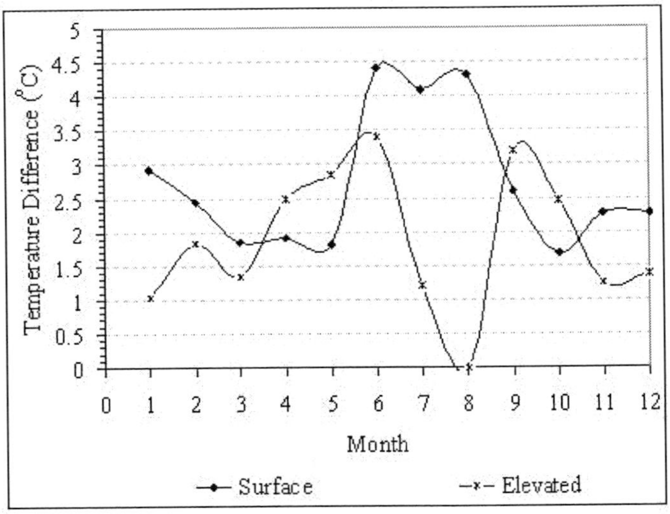

Figure 6. Average strength of the surface-based and elevated inversions in the various months (**Saudi Arabia Station Number 40416**)

4.1.4. Inversion strength

The average strength of the surface-based and elevated inversion layers in the various months of the years during the period of study in station 40416 is shown in Figure 6. The strength of the inversion or the thermal stability of the atmosphere in the present station was determined by calculating the temperature difference between the top and bottom of the inversion layer (ΔT). It can be seen from Figure 6 that the months of June, July and August tended to be the strongest in the surface-based inversions compared to the remaining months, which showed a drop in inversion strength.

The strength of the surface-based inversions was lower during March, April, May and October with an average around 1.7-1.9 °C. On the other hand, comparing the strength of the elevated inversions in the various months showed that during August, the temperature difference across the inversion layer was zero, and in the remaining months, it showed a slight increase in the inversion strength.

4.2. Analysis of thermal inversions in Saudi Arabia (Station NO. 40417)

4.2.1. Inversion frequencies

The results of the calculations of the inversion frequencies which occurred in the Kingdom of Saudi Arabia (Station Number 40417) are shown in Table 3. Figure 7 shows the frequency of occurrence of surface-based inversions, the frequency of elevated inversions, and the frequency of all inversions in this station. Figure 7 revealed that an inversion occurred in every month examined, except the months of April to October, which showed no occurrence of elevated inversions. Overall, most of the inversions tended to be surface based. Looking at the trend of the frequency of all inversions (F_{all}), it seems that the inversions are almost the same throughout the months. The maximum frequency of 50.0% occurred during February, while the minimum frequency of 45.6% occurred during April. It was also seen that the frequency of surface-based inversions ranged between 92.7% and 100%, and elevated inversions ranged between 1.6 and 7.3%.

Table 3. Monthly inversion frequencies (Saudi Arabia Station Number 40417) during 1994 - 2003

Month	N_{SBI}	No surface-based inversion	N_{elev}	No elevated inversion	N_{inv}	N_{sound}	F_{SBI}	F_{elev}	F_{all}
1	139	15	11	143	150	308	92.7	7.3	48.7
2	115	6	6	115	121	242	95.0	5.0	50.0
3	125	11	2	135	127	273	98.4	1.6	46.5
4	73	7	0	80	73	160	100.0	0.0	45.6
5	133	7	0	140	133	280	100.0	0.0	47.5
6	79	7	0	86	79	172	100.0	0.0	45.9
7	133	3	0	136	133	272	100.0	0.0	48.9
8	131	3	0	134	131	268	100.0	0.0	48.9
9	106	1	0	107	106	214	100.0	0.0	49.5
10	113	2	0	115	113	230	100.0	0.0	49.1
11	105	9	3	111	108	228	97.2	2.8	47.4
12	108	8	2	114	110	232	98.2	1.8	47.4

N_{SBI} is number of surface-based inversions, N_{elev} is number of elevated inversions, N_{inv} is total number of inversions, N_{sound} is number of soundings, F_{SBI} is frequency of occurrence of surface-based inversions, F_{elev} is frequency of occurrence of elevated inversions, and F_{all} is frequency of occurrence of all inversions.

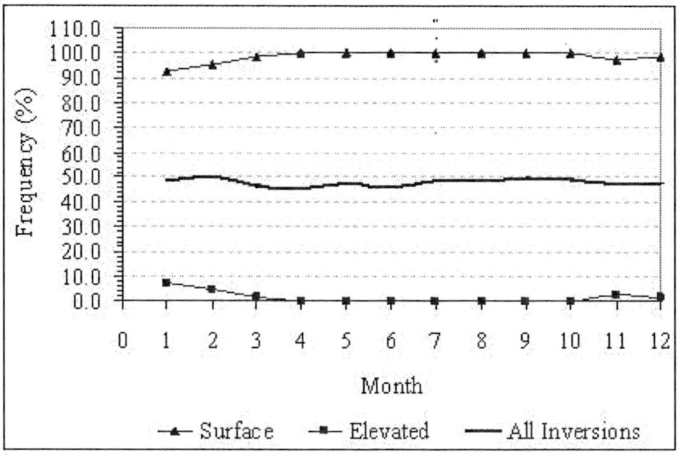

Figure 7. Monthly variations in inversion frequencies, given for each month the frequency of occurrence of surface-based inversions, the frequency of elevated inversions, and the frequency of all inversions (**Saudi Arabia Station Number 40417**)

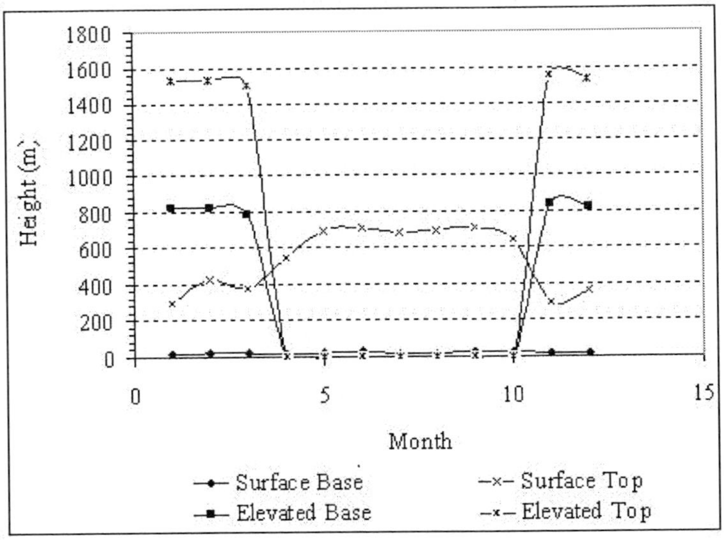

Figure 8. Mean height of the inversion base and top (m) according to the month of the year during the period of the study. Both surface-based and elevated inversions are included (**Saudi Arabia Station Number 40417**)

4.2.2. Inversion height

The mean height of the inversion base and top according to the month of the year during the period and area of the study in Station Number 40417 is shown in Figure 8. Both surface-based and elevated inversions are included in this Figure. The base of the surface-based inversion is defined by the station 40417 elevation with respect to the mean sea level. It can be seen that the heights of the top of the surface-based inversion layer reached the maximum depth during summer months, and almost the same as surface base during the remaining months of the year. On the other hand, the heights of the elevated inversions (top and base) were at their maximum during the months of January, February, March, November and December, and no elevated inversions occurred during the remaining months.

4.2.3. Inversion depth

The mean inversion depth was determined by calculating the difference in altitude between the top and the base of the inversion. The results of the mean inversion depth for both the surface-based and elevated inversions according to the month of the year during the period of the study in Saudi Arabia Station number 40417 are shown in Figure 9. The most obvious feature of Figure 9 was that the surface-based inversions were generally deeper than the elevated inversion layers during the summer months with an average values between 600 and 700 m. The reverse trend was true during the months of January, February, March, November and December, in which the elevated inversion layers were deeper than the surface-based inversion with average values between 700 and 720 m. The mean depths of the elevated inversions were zero during the remaining months.

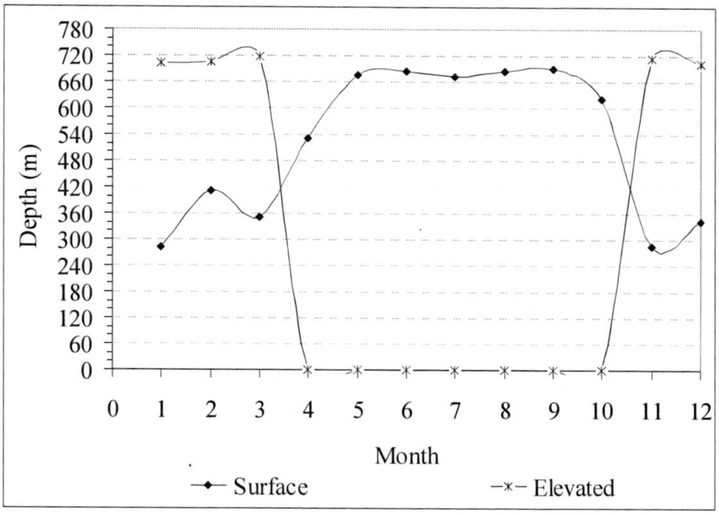

Figure 9. Mean inversion depth (m) according to the month of the year during the period of the study. Both surface-based and elevated inversions are included (Saudi Arabia Station Number 40417)

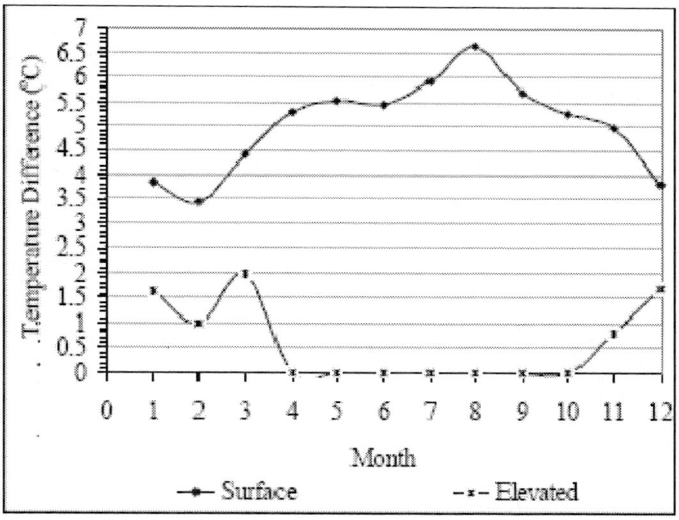

Figure 10. Average strength of the surface-based and elevated inversions in the various months (**Saudi Arabia Station Number 40417**)

4.2.4. Inversion strength

The average strengths of the surface-based and elevated inversion layer in the various months of the years during the period of study in Saudi Arabia Station Number 40417 are shown in Figure 10. The strength of the inversion or the thermal stability of the atmosphere in the present station was determined by calculating the temperature difference between the top and bottom of the inversion layer (ΔT). It can be seen from Figure 10 that the temperature difference was stronger in the surface-based inversions than the elevated inversion in all months. The strengths of the surface-based inversions generally increased from February until August, and then decreased throughout the remaining months. The average strengths of the surface-based inversion were in the range between 3.5 and 6.6 °C, with the maximum occurring during the month of August. On the other hand, the maximum strengths of the elevated inversion occurred during March with temperature difference of about 2 °C. The average strengths of the elevated inversions showed no temperature difference during April to October.

4.3. Analysis of thermal inversions in Saudi Arabia (Station NO. 41024)

4.3.1. Inversion frequencies

The results of the calculations of the inversion frequencies that occurred in the Kingdom of Saudi Arabia (Station Number 41024) are shown in Table 4. On the other hand, Figure 11 illustrates the frequency of occurrence of surface-based inversions, the frequency of elevated inversions, and the frequency of all inversions. Figure 11 reveals that an inversion occurred in every month examined, except the month of June and December, which showed no occurrence of elevated inversion. Overall, most of the inversions tended to be surface-based. From a close look at the curve of the frequency of all inversions (F_{all}), it would appear that the inversions were almost the same in all months, with the range between 42.6 and 48.9%. The maximum frequency of 48.9% occurred during June, and the minimum frequency of 42.6% occurred during February. In addition, June and December showed the frequencies of 100% in surface-based inversions, and the minimum of 94.5% occurred during September. The elevated inversion showed the opposite trend; i.e. the highest frequency of surface-based inversions in some months corresponded to the time of minimum elevated inversions in the others. The frequencies in the elevated inversion varied between 0.4 and 5.5%.

Table 4. Monthly inversion frequencies (Saudi Arabia Station Number 41024) during 1994 - 2003

Month	N_{SBI}	No surface-based inversion	N_{elev}	No elevated inversion	N_{inv}	N_{sound}	F_{SBI}	F_{elev}	F_{all}
1	272	27	5	294	277	598	98.2	1.8	46.3
2	201	42	7	238	208	488	96.6	3.4	42.6
3	217	47	11	254	228	529	95.2	4.8	43.1
4	198	29	6	223	204	456	97.1	2.9	44.7
5	243	15	5	257	248	520	98.0	2.0	47.7
6	170	4	0	174	170	348	100.0	0.0	48.9
7	241	10	1	250	242	502	99.6	0.4	48.2
8	232	28	1	262	233	523	99.6	0.4	44.6
9	241	21	14	249	255	525	94.5	5.5	48.6
10	254	11	2	265	256	532	99.2	0.8	48.1
11	230	25	1	261	231	517	99.6	0.4	44.7
12	242	24	0	268	242	534	100.0	0.0	45.3

$\mathbf{N_{SBI}}$ is number of surface-based inversions, $\mathbf{N_{elev}}$ is number of elevated inversions, $\mathbf{N_{inv}}$ is total number of inversions, $\mathbf{N_{sound}}$ is number of soundings, $\mathbf{F_{SBI}}$ is frequency of occurrence of surface-based inversions, $\mathbf{F_{elev}}$ is frequency of occurrence of elevated inversions, and $\mathbf{F_{all}}$ is frequency of occurrence of all inversions.

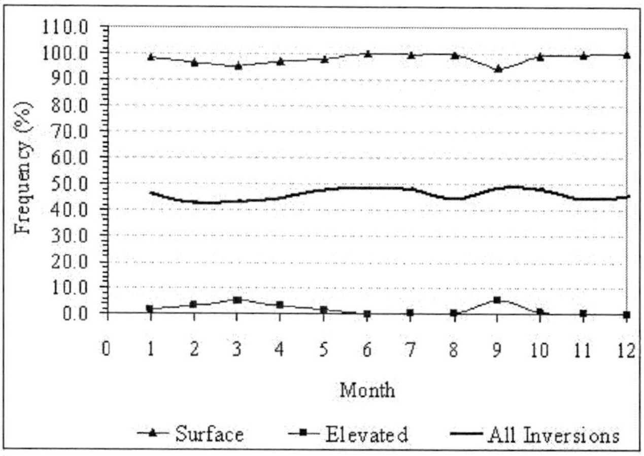

Figure 11. Monthly variations in inversion frequencies, given for each month the frequency of occurrence of surface-based inversions, the frequency of elevated inversions, and the frequency of all inversions (**Saudi Arabia Station Number 41024**)

4.3.2. Inversion height

The mean height of the inversion base and top according to the month of the year during the period and area of the study in Saudi Arabia Station Number 41024 is shown in Figure 12. Both bottom and top surface-based and elevated inversions are included in this Figure. The base of the surface-based inversion is defined by the station 41024 elevation with respect to the mean sea level. It can be seen that the mean heights of the surface base across surface-based inversion were almost constant throughout the months. The mean heights of the surface top increased from the first month up to June, after that they remained constant, and finally decreased after the month of September. On the other hand, the mean heights of the elevated bottom across the elevated inversion decreased from the first month up to June, increasing during July, August, October and November. The same trend was seen with the mean heights of the elevated top inversions.

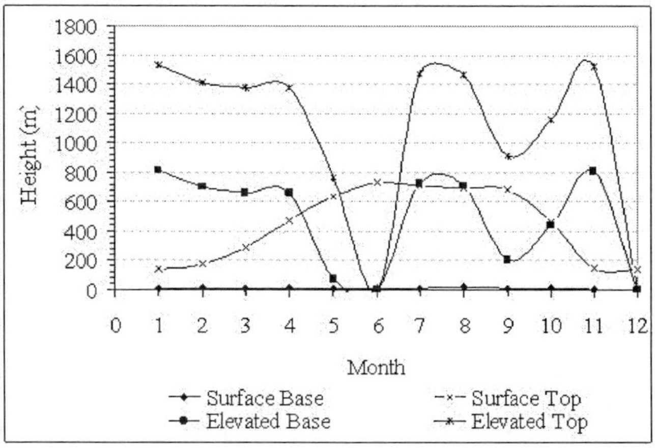

Figure 12. Mean height of the inversion base and top (m) according to the month of the year during the period of the study. Both surface-based and elevated inversions are included (**Saudi Arabia Station Number 41024**)

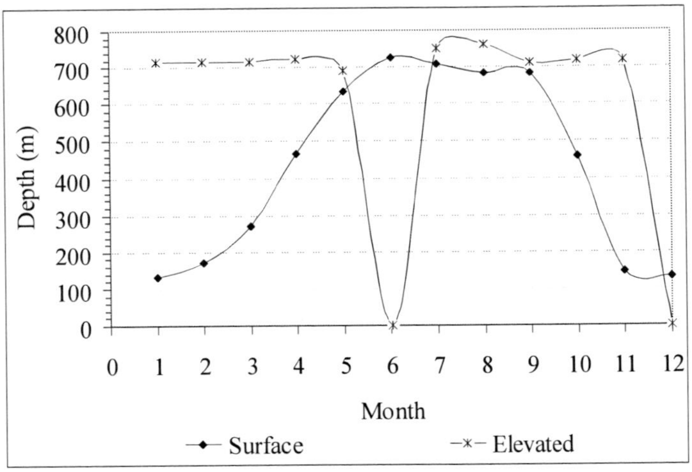

Figure 13. Mean inversion depth (m) according to the month of the year during the period of the study. Both surface-based and elevated inversions are included (**Saudi Arabia Station Number 41024**)

4.3.3. Inversion depth

The mean inversion depth was determined by calculating the difference in altitude between the top and the base of the inversion. The results of the mean inversion depth for both the surface-based and elevated inversions according to the month of the year during the period of the study in Saudi Arabia Station Number 41024 are shown in Figure 13. It can be seen that the elevated inversions were deeper than surface inversion during all months of the year, except the month of June in which no elevated inversion occurred. Moreover, the mean inversion depths between the surface and elevated inversions were high during the early months, and least during the last months of the year. For the surface-based inversion, the mean depth inversion reached the maximum values during June, July, August, September, and October, and the least in the remaining months. For the elevated inversion, the deepest inversions occurred in all months, except June and December, which showed zero elevated inversions.

4.3.4. Inversion strength

The average strength of the surface-based and elevated inversion layer in the various months of the years during the period of study in Saudi Arabia Station Number 41024 is shown in Figure 14. The strength of the inversion or the thermal stability of the atmosphere in the present station was determined by calculating the temperature difference between the top and bottom of the inversion layer (ΔT). Figure 14 indicates that all months showed higher temperature difference of the surface-based inversions compared to the elevated inversions. The strength of the surface-based inversions increased during the first months and reached the maximum during June with an average value of 6 °C. Thereafter, it decreased gradually towards the end of the year. On the other hand, the strength of the elevated inversions increased also during the first months, but decreased after reaching the maximum value of 2.8 °C during May. The strength of the elevated inversion increased again in the subsequent months up to September, and decreased toward the end of the year.

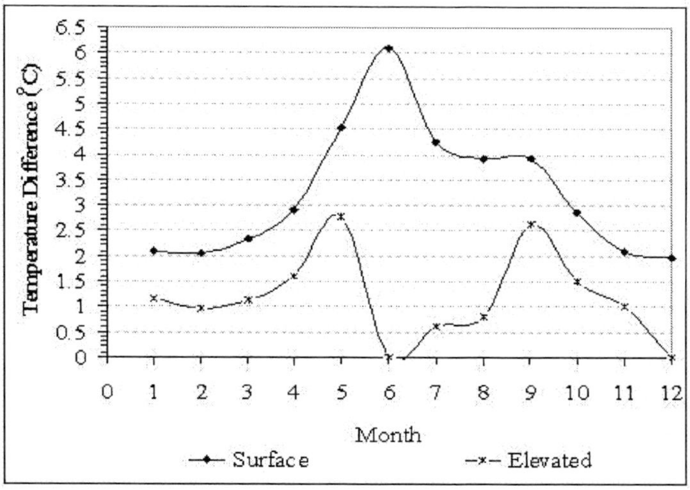

Figure 14. Average strength of the surface-based and elevated inversions in the various months (**Saudi Arabia Station Number 41024**)

Table 5. Monthly inversion frequencies (Kuwait Station Number 40582) during 1994 - 2003

Month	N_{SBI}	No surface-based inversion	N_{elev}	No elevated inversion	N_{inv}	N_{sound}	F_{SBI}	F_{elev}	F_{all}
1	208	57	149	117	357	531	58.3	41.7	67.2
2	15	9	14	10	29	48	51.7	48.3	60.4
3	188	67	127	127	315	509	59.7	40.3	61.9
4	23	4	13	14	36	54	63.9	36.1	66.7
5	210	61	73	197	283	541	74.2	25.8	52.3
6	NA	NA	NA	NA	NA	NA	NA	NA	NA
7	222	48	136	134	358	540	62.0	38.0	66.3
8	198	60	107	150	305	515	64.9	35.1	59.2
9	NA	NA	NA	NA	NA	NA	NA	NA	NA
10	226	41	100	167	326	534	69.3	30.7	61.0
11	NA	NA	NA	NA	NA	NA	NA	NA	NA
12	164	82	176	89	340	511	48.2	51.8	66.5

$\mathbf{N_{SBI}}$ is number of surface-based inversions, $\mathbf{N_{elev}}$ is number of elevated inversions, $\mathbf{N_{inv}}$ is total number of inversions, $\mathbf{N_{sound}}$ is number of soundings, $\mathbf{F_{SBI}}$ is frequency of occurrence of surface-based inversions, $\mathbf{F_{elev}}$ is frequency of occurrence of elevated inversions, and $\mathbf{F_{all}}$ is frequency of occurrence of all inversions.

4.4. Analysis of Thermal Inversions in Kuwait (Station NO. 40582)

4.4.1. Inversion frequencies

The results of the calculations of the inversion frequencies, which occurred in the State of Kuwait (Station Number 40582) are shown in Table 5. Figure 15 shows the frequency of occurrence of surface-based inversions, the frequency of elevated inversions, and the frequency of all inversions. Figure 15 reveals that an inversion occurred in every month examined, except the months of June, September, and November, which showed no occurrence of both surface-based and elevated inversions. Overall, surface-based inversion occurred more than elevated inversions in all months, except December, which showed high frequency in elevated inversions.

From an observation of the curve of the frequency of all inversions (F_{all}), it seems that the inversions had almost the same frequency in all months with the range between 52.3 and 67.2%. The maximum frequency of 67.2% occurred during January, and the minimum frequency of 52.3% occurred during May, and zero inversion occurred during June, September and November. Moreover, the month of May showed the highest frequencies of surface-based inversions compared to other months. The frequency of the elevated inversion was the highest during the month of December. In addition, the highest frequency of surface-based inversions in some months corresponded to the time of minimum elevated inversions in the others, except during December, which showed the highest frequency of elevated inversions.

4.4.2. Inversion height

The mean height of the inversion base and top according to the month of the year during the period and area of the study in Kuwait Station Number 40582 is shown in Figure 16. Both surface-based and elevated inversions are represented in Figure 16. The base of the surface-based inversion is defined by the station 40582 elevation with respect to the mean sea level. It can be seen that the heights of the elevated bottom and top were higher than surface base and top in all months of the year. The mean height of the surface base inversion was almost constant throughout the year. The heights of the surface top inversion were constant during the first months, but increased during July, August, and October. On the other hand, the elevated inversions were frequent in all months and their heights (top and base) were at their maximum during the month of July. By comparing the heights of the inversion top and base of both the surface-based and elevated inversions, it can be seen that the month of July indicated the highest inversion.

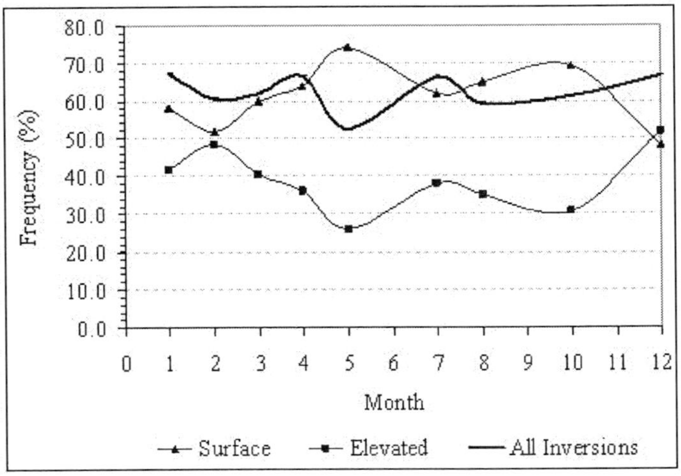

Figure 15. Monthly variations in inversion frequencies, given for each month the frequency of occurrence of surface-based inversions, the frequency of elevated inversions, and the frequency of all inversions (**Kuwait Station Number 40582**)

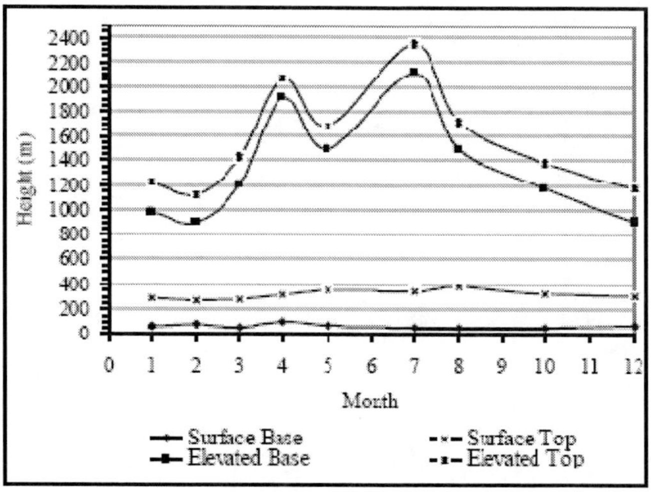

Figure 16. Mean height of the inversion base and top (m) according to the month of the year during the period of the study. Both surface-based and elevated inversions are included (**Kuwait Station Number 40582**)

4.4.3. Inversion depth

The results of the mean inversion depth for both the surface-based and elevated inversions according to the month of the year during the period of the study in Kuwait Station Number 40582 are shown in Figure 17. The surface-based inversions were generally deeper than the elevated inversion layers. It can also be seen that the depths of the elevated inversion were higher during the first months, but changed during the subsequent months. For the surface-based inversion, the mean depth inversion was maximum in August, and the least in the remaining months with an average around 200-300 m. For the elevated inversion, the deepest inversions occurred in December, and the least in the other months. The months of June, September and November showed no elevated inversion.

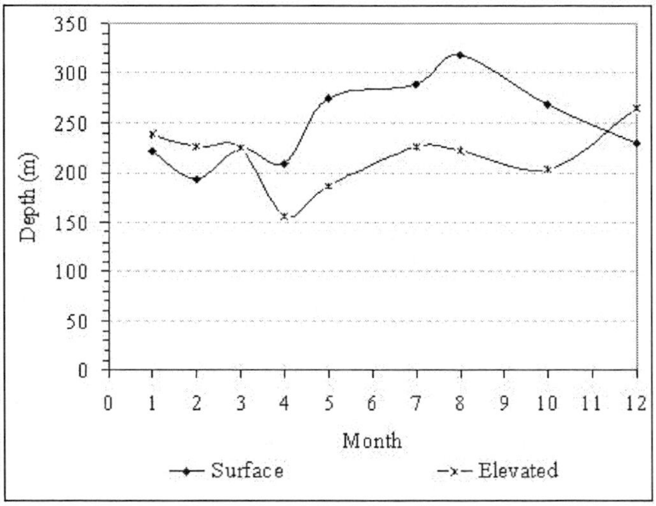

Figure 17. Mean inversion depth (m) according to the month of the year during the period of the study. Both surface-based and elevated inversions are included (**Kuwait Station Number 40582**)

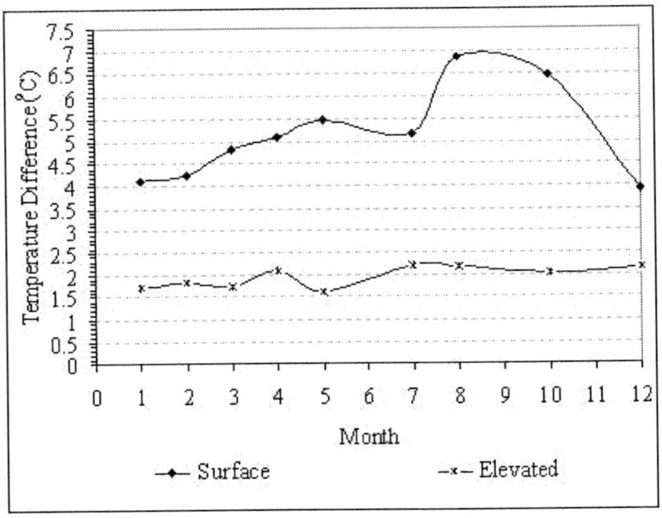

Figure 18. Average strength of the surface-based and elevated inversions in the various months (**Kuwait Station Number 40582**)

4.4.4. Inversion strength

The average strength of the surface-based and elevated inversion layer in the various months of the years during the period of study in Kuwait Station Number 40582 is shown in Figure 18. It can be seen from Figure 18 that the surface-based inversion was stronger than the elevated inversion in all months of the year. The month of August showed the maximum temperature difference across the surface-based inversion. The temperature difference across the surface-based inversion was high during the first months and deceased towards the end of the year. On the other hand, comparing the strength of the elevated inversions in the various months showed that during June, September and November, the temperature difference across the inversion layer was zero. The maximum strength of the elevated inversion occurred during July and August.

4.5. Analysis of Thermal Inversions in the United Arab Emirates (Station NO. 41217)

4.5.1. Inversion frequencies

The results of the calculations of the inversion frequencies, which were recorded in the United Arab Emirates (Abu Dhabi Station NO. 41217) are shown in Table 6. Figure 19 shows the frequency of occurrence of surface-based inversions, the frequency of elevated inversions, and the frequency of all inversions. Figure 19 reveals that an inversion occurred in every month examined, except the month of May and July, which showed no occurrence of elevated inversion. Overall, most of the inversions tended to be surface-based. Figure 19 also indicates an increase in the frequency of surface-based inversion up to month of July followed by a decrease in the frequency toward the end of the year. The reverse trend was true for the elevated inversion, where the frequency decreased up to the month of May followed by an increase in the frequency towards the end of the year. From a close look at the curve of the frequency of all inversions (F_{all}), it appears that the frequency of the inversions generally increased toward the end of the year. The maximum frequency of 57.1% occurred during December, and minimum frequency of 38.5% occurred during August. Since F_{elev} is the complement of F_{SBI}, an opposite trend in elevated inversions was observed. In addition, it can be seen that the highest frequency of surface-based inversions in some months corresponded to the time of minimum elevated inversions in the others.

Table 6. Monthly inversion frequencies (Abu Dhabi Station Number 41217) during 1994 - 2003

Month	N_{SBI}	No surface-based inversion	N_{elev}	No elevated inversion	N_{inv}	N_{sound}	F_{SBI}	F_{elev}	F_{all}
1	255	24	51	228	306	558	83.3	16.7	54.8
2	213	46	30	229	243	518	87.7	12.3	46.9
3	225	54	14	265	239	558	94.1	5.9	42.8
4	247	23	6	264	253	540	97.6	2.4	46.9
5	268	11	0	279	268	558	100.0	0.0	48.0
6	209	37	1	269	210	516	99.5	0.5	40.7
7	229	45	0	279	229	553	100.0	0.0	41.4
8	213	66	2	277	215	558	99.1	0.9	38.5
9	221	47	3	264	224	535	98.7	1.3	41.9
10	254	9	7	256	261	526	97.3	2.7	49.6
11	260	10	20	250	280	540	92.9	7.1	51.9
12	230	18	54	195	284	497	81.0	19.0	57.1

$\mathbf{N_{SBI}}$ is number of surface-based inversions, $\mathbf{N_{elev}}$ is number of elevated inversions, $\mathbf{N_{inv}}$ is total number of inversions, $\mathbf{N_{sound}}$ is number of soundings, $\mathbf{F_{SBI}}$ is frequency of occurrence of surface-based inversions, $\mathbf{F_{elev}}$ is frequency of occurrence of elevated inversions, and $\mathbf{F_{all}}$ is frequency of occurrence of all inversions.

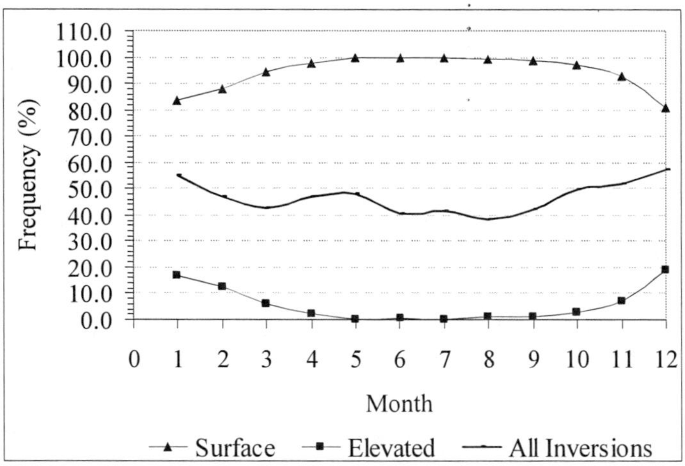

Figure 19. Monthly variations in inversion frequencies, given for each month, the frequency of occurrence of surface-based inversions, the frequency of elevated inversions, and the frequency of all inversions (**Abu Dhabi Station Number 41217**)

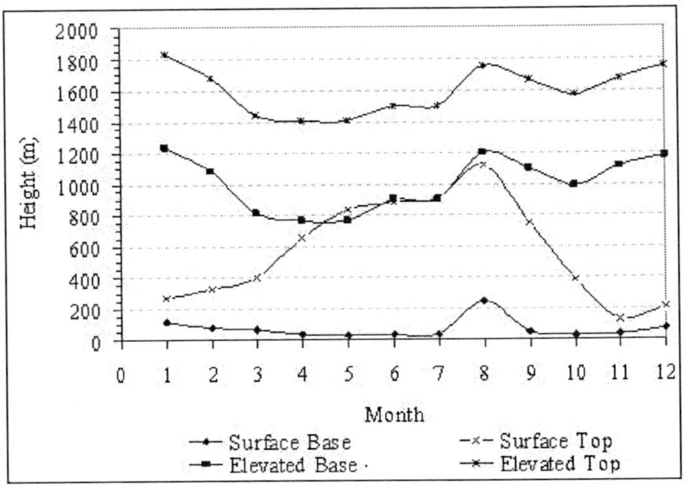

Figure 20. Mean height of the inversion base and top (m) according to the month of the year during the period of the study. Both surface-based and elevated inversions are included (**Abu Dhabi Station Number 41217**)

4.5.2. Inversion height

The mean height of the inversion base and top according to the month of the year during the period and area of the study in Abu Dhabi Station Number 41217 is shown in Figure 20. Both surface-based and elevated inversions are included in this Figure. The base of the surface-based inversion is defined by station 41217 elevation with respect to the mean sea level. It can be seen that the heights of the surface base were almost constant throughout the year. The mean heights of the surface top were constant in the beginning of the year, and then increased during July, August, October, and December. The maximum heights of the surface top occurred during August. On the other hand, the elevated inversions were frequent in all months and their heights (top and base) were at their maximum during the month of July. By comparing the heights of the inversion top and base of both the surface-based and elevated inversions, it can be seen that the month of July indicated the highest inversion.

4.5.3. Inversion depth

The results of the mean inversion depth for both the surface-based and elevated inversions according to the month of the year during the period of the study in Abu Dhabi Station Number 41217 are shown in Figure 21. The most obvious feature of Figure 21 was that the surface-based inversion curve took a bell shape, with the depth of the inversion increasing from the beginning of the year up to month of July, and then gradually decreasing toward the end of the year. The maximum depth of the surface-based inversion with an average of about 880 m occurred during July. The depths of the elevated inversion were almost constant throughout the year with an average depth of around 550-620 m. No inversion occurred during May and July. Overall, the elevated inversions were deeper than the surface-based inversion layers during the winter months. However, the surface based inversion layers were deeper during the summer months.

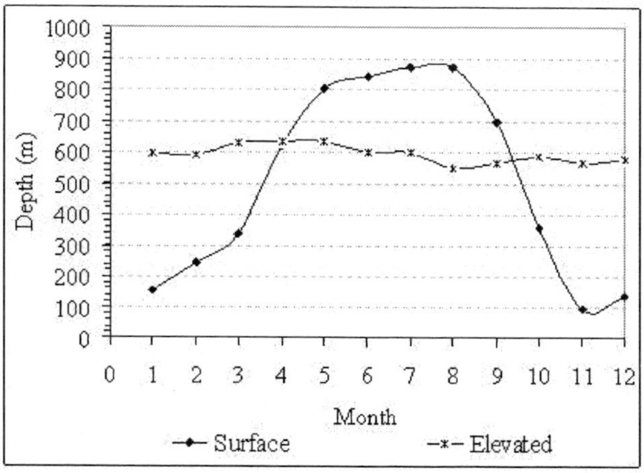

Figure 21. Mean inversion depth (m) according to the month of the year during the period of the study. Both surface-based and elevated inversions are included (**Abu Dhabi Station Number 41217**)

4.5.4. Inversion strength

The average strength of the surface-based and elevated inversion layer in the various months of the years during the period of the study in Abu Dhabi Station Number 41217 is shown in Figure 22. It can be generally seen that the average strengths of the surface-based inversion increased with the months of the year, whereas the strengths of the elevated inversion decreased over months. The temperature difference of the surface-based inversion generally increased from January up to June, and then decreased toward the end of the year. The strength of the elevated inversion showed a slight increase in temperature difference during July compared to other months. The maximum strength of approximately 9.4 °C of the surface-based inversion occurred during June, whereas the maximum strength of around 3.7 °C of the elevated inversion occurred during January followed by July, which is around 3.4 °C.

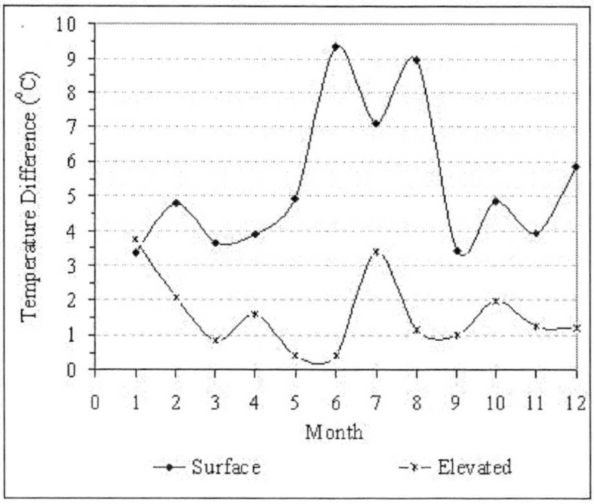

Figure 22. Average strength of the surface-based and elevated inversions in the various months (**Abu Dhabi Station Number 41217**)

4.6. Analysis of Thermal Inversions in Oman (Salalah Station NO. 41316)

4.6.1. Inversion frequencies

The results of the calculations of the inversion frequencies occurred in Oman (Salalah Station Number 41316) are shown in Table 7. Figure 23 shows the frequency of the occurrence of surface-based inversions, the frequency of elevated inversions, and the frequency of all inversions. Figure 23 reveals that an inversion occurred in every month examined, except February, April, June, September and November, which showed no occurrence of both surface-based and elevated inversions. Overall, the elevated inversions occurred more than surface-based inversion in all months. The curve of the frequency of all inversions showed that the inversions reached the maximum frequency of 48.7% during January, and its minimum of 34.4% during March. Moreover, July and August showed the highest frequencies of elevated inversions, 99.0 and 99.2%, respectively, compared to the other months. The highest frequency of surface-based inversions (26.6%) occurred during January and the minimum (1.0%) occurred during July. In general, the highest frequency of elevated inversions in all months corresponded to the time of minimum surface-based inversions in those months.

Table 7. Monthly Inversion Frequencies (Oman Station Number 41316) during 1994 - 2003

Month	N_{SBI}	No surface-based inversion	N_{elev}	No elevated inversion	N_{inv}	N_{sound}	F_{SBI}	F_{elev}	F_{all}
1	41	117	113	45	154	316	26.6	73.4	48.7
2	NA	NA	NA	NA	NA	NA	NA	NA	NA
3	23	144	92	75	115	334	20.0	80.0	34.4
4	NA	NA	NA	NA	NA	NA	NA	NA	NA
5	9	122	109	22	118	262	7.6	92.4	45.0
6	NA	NA	NA	NA	NA	NA	NA	NA	NA
7	1	101	95	7	96	204	1.0	99.0	47.1
8	1	145	127	19	128	292	0.8	99.2	43.8
9	NA	NA	NA	NA	NA	NA	NA	NA	NA
10	27	127	112	42	139	308	19.4	80.6	45.1
11	NA	NA	NA	NA	NA	NA	NA	NA	NA
12	32	142	116	58	148	348	21.6	78.4	42.5

$\mathbf{N_{SBI}}$ is number of surface-based inversions, $\mathbf{N_{elev}}$ is number of elevated inversions, $\mathbf{N_{inv}}$ is total number of inversions, $\mathbf{N_{sound}}$ is number of soundings, $\mathbf{F_{SBI}}$ is frequency of occurrence of surface-based inversions, $\mathbf{F_{elev}}$ is frequency of occurrence of elevated inversions, and $\mathbf{F_{all}}$ is frequency of occurrence of all inversions.

NA: Not Available

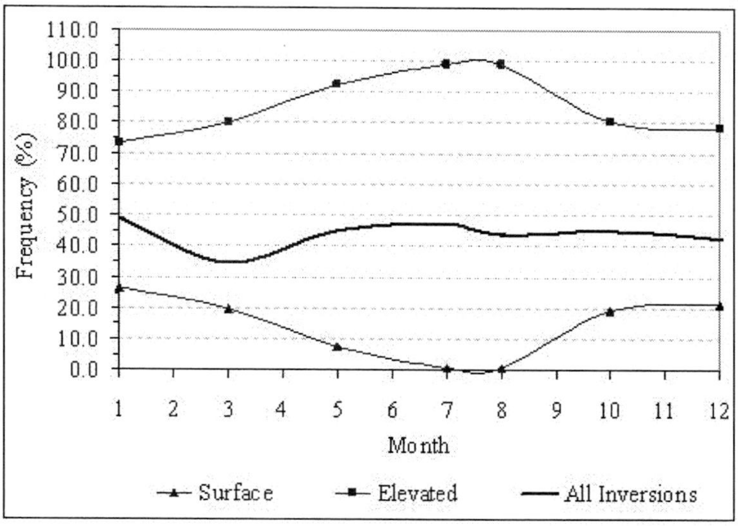

Figure 23. Monthly variations in inversion frequencies, given for each month the frequency of occurrence of surface-based inversions, the frequency of elevated inversions, and the frequency of all inversions (**Oman Station Number 41316**).

4.6.2. Inversion height

The mean height of the inversion base and top according to the month of the year during the period and area of the study in Oman Station Number 41316 is shown in Figure 24. Both surface-based and elevated inversions are included in Figure 24. The base of the surface-based inversion is defined by the station 41316 elevation with respect to the mean sea level. Figure 24 shows that the mean heights of the elevated bottom and top were higher than surface base and top in all months of the years. The mean height of the surface base inversion was almost constant through out the year, except during May, which showed an increase in the height of the surface base inversion. The heights of the surface top inversion increased during the first months, followed by a decrease in the inversion, and finally increased towards the end of the year.

On the other hand, the mean heights of the elevated inversions bottom and top showed a decrease in the inversions during the first months, followed by an increase in the inversion towards the end of the period of the study. By comparing the mean heights of the inversion bottom and top of both the surface-based and elevated inversions, it can be seen that the inversion heights generally decreased during the first months of the year, and increased towards the end of the period of the study.

4.6.3. Inversion depth

The results of the mean inversion depth for both the surface-based and elevated inversion according to the months of the year during the period of the study in Salalah Station Number 41316 are shown in Figure 25. This Figure shows that the elevated inversion layers were generally deeper than the surface-based inversion. In addition, the depth of the elevated inversion was higher than that of surface-based inversion during all months. The mean depth of the surface-based inversion reached its maximum during May, and the minimum during August. For the elevated inversion, the deepest inversion occurred during August, and the least inversion occurred during October. No surface-based and elevated inversions were recorded during February, April, June, September and November.

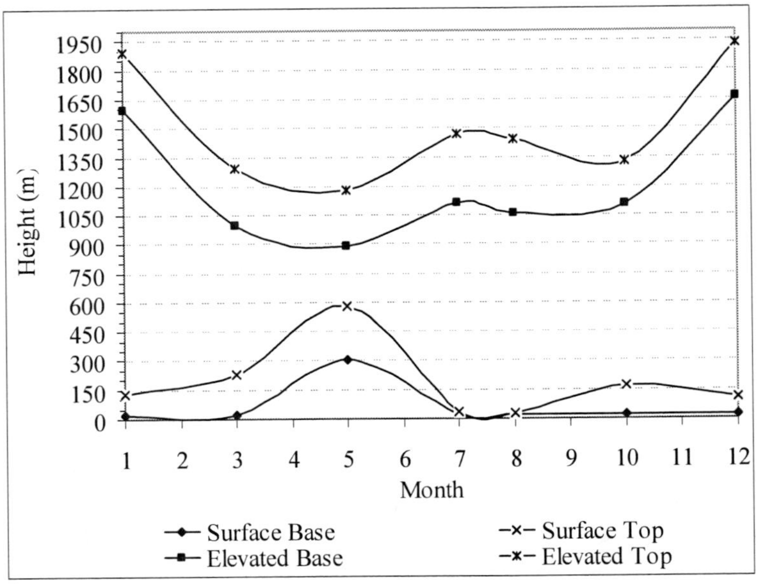

Figure 24. Mean height of the inversion base and top (m) according to the month of the year during the period of the study. Both surface-based and elevated inversions are included (**Oman Station Number 41316**)

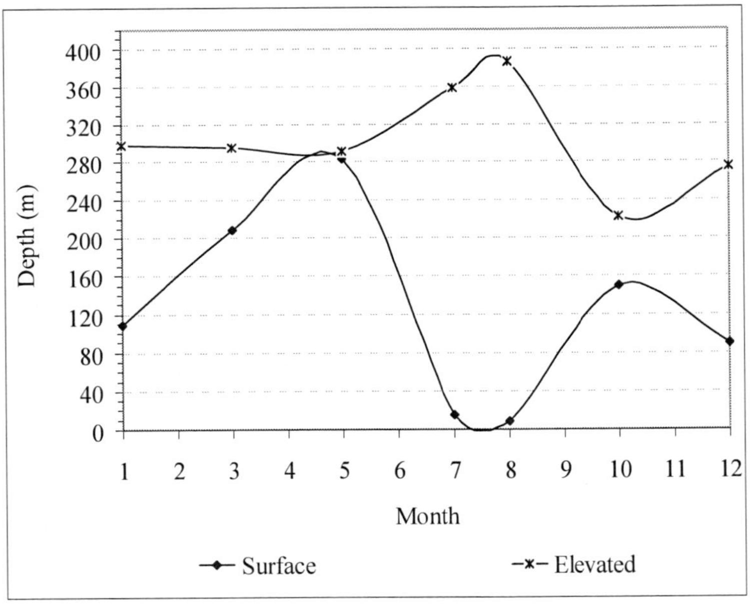

Figure 25. Mean inversion depth (m) according to the month of the year during the period of the study. Both surface-based and elevated inversions are included (**Oman Station Number 41316**)

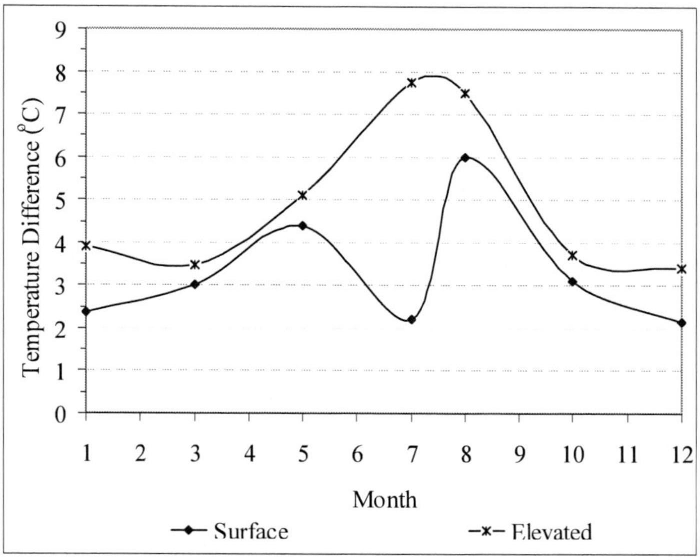

Figure 26. Average strength of the surface-based and elevated inversions in the various months (**Oman Station Number 41316**)

4.6.4. Inversion strength

The average strength of the surface-based and elevated inversion layers in the various months of the years during the period of study in Salalah Station 41316 is shown in Figure 26. This Figure revealed that the elevated inversion was stronger than the surface-based inversions in all months of the year. The maximum temperature difference across the elevated inversion occurred during August, and the minimum occurred during December. The temperature difference across the surface-based and elevated inversions generally increased through out the months of the year, followed by a decrease in the temperature towards the end of the study.

5. Discussion and Conclusion

A general comparison of data from the stations involved in the study areas indicated monthly changes in inversion frequency, height, depth, and strength. The frequency of the occurrence of all inversions seemed to be slightly stable throughout the period of the study in all stations. Since the frequency of the elevated inversions is the complement of the surface-based inversions, the highest frequency of surface-based inversions in some months corresponded to the time of minimum elevated inversions in the others. Overall, most of the inversions tended to be surface-based in all stations of the study areas in the Gulf Countries. The surface-based and elevated inversions vary from winter to summer months depending on the changes of the atmospheric weather parameters in all stations of the study areas. The elevated inversions were generally lower and weaker than the surface-based inversions. The results of this study showed some similarities with the previous results reported by Abdul-Wahab et al. (2004), Abdul-Wahab (2003), Kahl et al. (1993), and Kahl (1990).

A further comparison of the frequency of monthly inversions in all stations in the study areas showed that the frequency of thermal inversions at the three stations in Saudi Arabia (Stations No. 40416, 40417, 41024) tended to be surface-based. The frequency of surface-based inversions at Stations No. 40417 and 41024 was higher than that in station No. 40416. However, the frequency of elevated inversion at station No. 40416 was higher in all months of the year compared to stations No. 40417 and 41024. A similar trend was observed with the occurrence of all inversions at these stations.

Compared with the three stations in Saudi Arabia, the frequency of occurrence of surface-based inversion at Kuwait station No. 40582 was not as high as in Saudi Arabia. The occurrence of elevated inversion at the Kuwait station was higher than that in Saudi Arabia. The monthly inversion frequencies at Abu Dhabi station No. 41217 followed the same trends as those stations in Saudi Arabia, especially stations No. 40417 and 41024. Different trends of the occurrence of surface-based and elevated inversions were observed at Salalah station No. 41316. The occurrence of elevated inversions at this station was higher than the surface-based inversion. Therefore, the occurrence of elevated inversion was higher than that in other stations of study areas, and the occurrence of surface-based inversion was lower throughout the months of the year compared with other stations.

Comparing the inversion height of both surface and elevated base and top, the surface-based inversion in all stations of study areas was stable throughout the months of the year. The height of surface top inversion at Saudi Arabia station No. 40416 was stable throughout the months. However, the height of the surface top inversion at Saudi Arabia stations No. 40417 and 41024 increased during summer months and decreased during winter. On the other hand, the heights of elevated base and top inversion generally decreased throughout the months of the year, with an increase during July. The study also indicated annual variations in the thermal inversions associated with inversion depth and height; at Saudi Arabia station No. 40416. In contrast, the heights of the elevated base and top inversion were high during winter months and low during summer months at Saudi Arabia stations No. 40417 and 41024.

Compared to the Saudi Arabia stations, different scenarios of inversion height were observed at Kuwait station No. 40582, Abu Dhabi station No. 41217, and Salalah station No. 41316. The heights of both surface-based and elevated inversions (base and top) followed almost the same trend at Kuwait station No. 40582 and Abu Dhabi station No. 41217. The heights of the surface base and top at these two stations were stable throughout the months of the year with a decrease in the inversion heights during June, September, and November. The heights of the elevated base and top at both stations generally decreased throughout the months of the year with an increase during April, July, and October. For Salalah station No. 41316, the heights of the elevated bottom and top were higher than surface base and top in all months of the year compared with other station. In addition, this station indicated that the inversion heights generally decreased during the first month of the year, and increased toward the end of the study.

The mean inversion depths for both the surface-based and elevated inversions followed different trends at the three stations in Saudi Arabia stations No. 40416, 40417, and 41024. The depths of surface-based inversions were high during summer months and low during winter at the three stations. On the other hand, the depths of elevated inversions generally decreased throughout the months of the year at these stations. The depths of surface-based and elevated inversions fluctuated at Kuwait station No. 40582. The depths of surface-based inversions were high during summer months and low during winter at Abu Dhabi station No.

41217. However, the depths of elevated inversions at this station were generally stable throughout the year, with no elevated inversion occurred during May and July. In contrast, Salalah station No. 41316 revealed that the elevated inversion layers were generally deeper than the surface-based inversion. Compared with other stations in the study areas, the depths of the elevated inversion were higher than that of surface-based inversions during all months of the year.

The temperature difference across the surface-based inversion was higher during summer months and low during winter at all stations of the study area. However, the temperature difference across the elevated inversions was generally low during summer months and high during winter at all stations, except Kuwait station No. 40582, which showed an increase in inversion strength during July and August. An increase in inversion strength was also observed during July at Abu Dhabi station No. 41217. In Salalah station No. 41316, the elevated inversion was stronger than the surface-based inversions in all months of the year. This station also indicated that the temperature difference across the surface-based inversions generally increased throughout the months of the year, followed by a decrease in temperature toward the end of the study.

The annual inversions throughout the period of the study were compared between some of the selected stations in the study areas including Saudi Arabia station No. 41024, Kuwait station No. 40582, Abu Dhabi station No. 41217, and Salalah station No. 41316. The results indicated annual variations in the thermal inversions associated with inversion depth, height, temperature and strength across both the surface-based and elevated inversions throughout the 10-year period.

References

Abdul-Wahab, S., Bouhamra, W., Ettouney, H., Sowerby, B. & Crittenden, B. (1999). Prediction of air pollution around heavily industrialised areas: Use of the industrial source complex short-term model with emissions from a large number of sources. *Environmental Protection Bulletin*, *58*, 3-13.

Abdul-Wahab, S. A. & Al-Alawi, S. M. (2002). Assessment and prediction of tropospheric ozone concentration levels using artificial neural networks. *Environmental Modelling and Software*, *17(3)*, 219-228.

Abdul-Wahab, S. A. (2003). Analysis of thermal inversions in the Khareef Salalah region in the Sultanate of Oman. *Journal of Geophysical Research D: Atmospheres*, *108(9)*, ACL 6-1 ACL 6-8.

Abdul-Wahab, S. A., Al-Saifi, S. Y., Alrumhi, B. A., Abdulraheem, M. Y. & Al-Uraimi, M. (2004). Determination of the features of the low-level temperature inversions above a suburban site in Oman using radiosonde temperature measurements: Long-term analysis. *Journal of Geophysical Research D: Atmospheres*, *109(20)*, D20101 1-9.

Abdul-Wahab, S. A., Bakheit, C. S. & Siddiqui, R. A. (2005). Study the relationship between the health effects and characterization of thermal inversions in the Sultanate of Oman. *Atmospheric Environment*, *39(30)*, 5466-5471.

AeroVironment, Inc. (1976). The monostatic acoustic radar, *Monrovia*, Calif.

Aikawa, M. & Hiraki, T. (2009). Characteristic seasonal variation of vertical air temperature profile in urban areas of Japan. *Meteorology and Atmospheric Physics, 104(1-2)*, 95-102.

Angell, J. K. & Korshover, J. (1983). Global temperature variations in the troposphere and stratosphere, 1958-1982. *Monthly Weather Review, 111(5)*, 901-921.

Angell, J. K. (1988). Variations and trends in tropospheric and stratospheric global temperatures, 1958-1987. *Journal of Climate, 1*, 1296-1313.

Anquetin, S., Guilbaud, C. & Chollet, J. P. (1998). The formation and destruction of inversion layers within a deep valley. *Journal of Applied Meteorology, 37(12)*, 1547-1560.

Ballish, B. A. & Kumar, V. K. (2008). Systematic differences in aircraft and radiosonde temperatures. *Bulletin of the American Meteorological Society, 89(11)*, 1689-1707.

Bilello, M. A. (1966). Survey of Arctic and subarctic temperature inversions. U.S. *Army Cold Regions Research & Engineering Laboratory*, Hanover, N.H., *1-36*. TR 161.

Bourne, S. M., Bhatt, U. S., Zhang, J. & Thoman, R. (2010). Surface-based temperature inversions in Alaska from a climate perspective. *Atmospheric Research, 95(2-3)*, 353-366.

Bradley, R. S., Keimig, F. T. & Diaz, H. F. (1992). Climatology of surface-based inversions in the North American Arctic. *Journal of Geophysical Research, 97(D14)*, 15,699-712.

Bradley, R. S., Keimig, F. T. & Diaz, H. F. (1993). Recent changes in the North American Arctic boundary layer in winter, *Journal of Geophysical Research, 98(D5)*, 8851-8858.

Busch, N., Ebel, U., Kraus, H. & Schaller, E. (1982). The structure of the subpolar inversion-capped ABL. *Archives for Meteorology, Geophysics, and Bioclimatology Series A, 31(1-2)*, 1-18.

Chang, T. Y., Norbeck, J. M. & Weinstock, B. (1980). NO_2 air quality-precursor relationship: An ambient air quality evaluation in the Los Angeles Basin. *Journal of the Air Pollution Control Association, 30(2)*, 157-162.

Chen, W. N., Tsao, C. C. & Nee, J. B. (2004). Rayleigh lidar temperature measurements in the upper troposphere and lower stratosphere. *Journal of Atmospheric and Solar-Terrestrial Physics, 66(1)*, 39-49.

Chutko, K. J. & Lamoureux, S. F. (2009). The influence of low-level thermal inversions on estimated melt-season characteristics in the central Canadian Arctic. *International Journal of Climatology, 29(2)*, 259-268.

Devara, P. C. S., Raj, P. E. & Sharma, S. (1994). Remote sensing of atmospheric aerosol in the nocturnal boundary layer using lidar. *Environmental Pollution, 85(1)*, 97-102.

Devasthale, A., Willén, U., Karlsson, K. G. & Jones, C. G. (2010). Quantifying the clear-sky temperature inversion frequency and strength over the Arctic Ocean during summer and winter seasons from AIRS profiles, *Atmospheric Chemistry and Physics Discussions, 10(2)*, 2835-2858.

Divakarla, M. G., Barnet, C. D., Goldberg, M. D., Maddy, E., Wolf, W., Flynn, L., Xiong, X., Wei, J., Zhou, L. & Xingpinliu, (2006). Validation of Atmospheric Infrared Sounder (AIRS) temperature, water vapor and ozone retrievals with matched radiosonde and ozonesonde measurements and forecasts. *Proceedings of SPIE - The International Society for Optical Engineering, 6405*, Article number 640503.

El-Askary, H. & Kafatos, M. (2008). Investigation of thermal inversions as a major contributer to the black cloud episodes over Cairo. *International Geoscience and Remote Sensing Symposium (IGARSS)* , art. no. 4423802, 4298-4301.

El Desouky, M. & Abdul-Wahab, S. A. (1988). Ground based thermal inversion in Kuwait, *Environ. Prot. Dep.*, Kuwait.

Gaffen, D. J., Elliott, W. P. & Robock, A. (1992a). Relationships between tropospheric water vapor and surface temperature as observed by radiosondes. *Geophysical Research Letters, 19(18)*, 1839-1842.

Gaffen, D. J., Robock, A. & Elliott, W. P. (1992b). Annual cycles of tropospheric water vapour. *Journal of Geophysical Research, 97(D16)*, 18, 185-193.

Gaffen, D. J. (1994). Temporal inhomogeneities in radiosonde temperature records. *Journal of Geophysical Research, 99(D2)*, 3667-3676.

Galerkin, L. I., Gritsenko, A. M. & Panfilova, S. G. (1998). Climatic characteristics of the layer of thermal inversions in the Ocean. *Oceanology, 38(6)*, 758-765.

Gopalakrishna, V. V., Johnson, Z., Salgaonkar, G., Nisha, K., Rajan, C. K. & Rao, R. R. (2005). Observed variability of sea surface salinity and thermal inversions in the Lakshadweep Sea during contrast monsoons. *Geophysical Research Letters, 32(18)*, art. no. L18605, 1-4.

Grant, A. N., Brönnimann, S., Ewen, T. & Nagurny, A. (2009). A new look at radiosonde data prior to 1958. *Journal of Climate, 22(12)*, 3232-3247.

Haagenson, P. L. (1979). Meteorological and climatological factors affecting Denver air quality. *Atmospheric Environment - Part A General Topics, 13(1)*, 79-85.

Haimberger, L. (2007). Homogenization of radiosonde temperature time series using innovation statistics. *Journal of Climate, 20(7)*, 1377-1403.

Hosler, C. R. (1961). Low level inversion frequency in the contiguous United States. *Monthly Weather Review, 59*, 319-339.

Hudson, S. R., Town, M. S., Walden, V. P. & Warren, S. G. (2004). Temperature, humidity, and pressure response of radiosondes at low temperatures. *Journal of Atmospheric and Oceanic Technology, 21(5)*, 825-836.

Hudson, S. R. & Brandt, R. E. (2005). A look at the surface-based temperature inversion on the Antarctic Plateau. *Journal of Climate, 18(11)*, 1673-1696.

Janhäll, S., Olofson, K. F. G., Andersson, P. U., Pettersson, J. B. C. & Hallquist, M. (2006). Evolution of the urban aerosol during winter temperature inversion episodes. *Atmospheric Environment, 40(28)*, 5355-5366.

Kahl, J. D. (1990). Characteristics of the low-level temperature inversion along the Alaskan Arctic coast. *International Journal of Climatology, 10(5)*, 537-548.

Kahl, J. D., Serreze, M. C. & Schnell, R. C. (1992). Tropospheric low-level temperature inversions in the Canadian Arctic. *Atmosphere Ocean, 30(4)*, 511-529.

Kahl, J. D. W., Serreze, M. C., Stone, R. S., Shiotani, S., Kisley, M. & Schnell, R. C. (1993). Tropospheric temperature trends In the Arctic: 1958-1986. *Journal of Geophysical Research, 98(D7)*, 12, 825-838.

Kahl, J. D. W., Martinez, D. A. & Zaitseva, N. A. (1996). Long-term variability in the low-level inversion layer over the Arctic Ocean. *International Journal of Climatology, 16(11)*, 1297-1313.

Karoly, D. J. (1989). Northern Hemisphere temperature trends: a possible greenhouse gas effect. *Geophysical Research Letters, 16(5)*, 465-468.

Kassomenos, P. A. & Koletsis, I. G. (2005). Seasonal variation of the temperature inversions over Athens, Greece. *International Journal of Climatology, 25(12)*, 1651-1663.

Kelly, P. M., Jones, P. D., Sear, C. B., Cherry, B. S. G. & Tavakol, R. K. (1982). Variations in surface air temperatures: part 2, Arctic regions, 1881-1980. *Monthly Weather Review, 110(2)*, 71-83.

Kukkonen, J., Pohjola, M., Sokhi, R. S., Luhana, L., Kitwiroon, N., Fragkou, L., Rantamäki, M.., Berge, E., Odegaard, V., Slordal, L. H., Denby, B. & Finardi, S. (2005). Analysis and evaluation of selected local-scale PM10 air pollution episodes in four European cities: Helsinki, London, Milan and Oslo. *Atmospheric Environment, 39(15)*, 2759-2773.

Kurian, J. & Vinayachandran, P. N. (2006). Formation mechanisms of temperature inversions in the southeastern Arabian Sea. *Geophysical Research Letters, 33(17)*, art. no. L17611.

Liu, Y. & Key, J. R. (2003). Detection and analysis of clear-sky, low-level atmospheric temperature inversions with MODIS. *Journal of Atmospheric and Oceanic Technology, 20(12)*, 1727-1737.

Luers, J. K. & Eskridge, R. E. (1998). Use of radiosonde temperature data in climate studies. *Journal of Climate, 11(5)*, 1002-1019.

Mahesh, A., Walden, V. P. & Warren, S. G. (1997). Radiosonde temperature measurements in strong inversions: Correction for thermal lag based on an experiment at the South Pole. *Journal of Atmospheric and Oceanic Technology, 14(1)*, 45-53.

Majorowicz, J. A. & Safanda, J. (1998). Ground surface temperature history from inversions of underground temperatures - a case study of the Western Canadian Sedimentary Basin. *Tectonophysics, 291(1-4)*, 287-298.

Malek, E., Davis, T., Martin, R. S. & Silva, P. J. (2006). Meteorological and environmental aspects of one of the worst national air pollution episodes (January, 2004) in Logan, Cache Valley, Utah, USA. *Atmospheric Research, 79(2)*, 108-122.

Milionis, A. E. & Davies, T. D. (1992). A five-year climatology of elevated inversions at Hemsby (UK). *International Journal of Climatology, 12(2)*, 205-215.

Milionis, A. E. & Davies, T. D. (2002). Associations between atmospheric temperature inversions and vertical wind profiles: A preliminary assessment. *Meteorological Applications, 9(2)*, 223-228.

Milionis, A. E. & Davies, T. D. (2008a). A comparison of temperature inversion statistics at a coastal and a non-coastal location influenced by the same synoptic regime. *Theoretical and Applied Climatology, 94(3-4)*, 225-239.

Milionis, A. E. & Davies, T. D. (2008b). The effect of the prevailing weather on the statistics of atmospheric temperature inversions. *International Journal of Climatology, 28(10)*, 1385-1397.

Nisha, K., Rao, S. A., Gopalakrishna, V. V., Rao, R. R., Girishkumar, M. S., Pankajakshan, T., Ravichandran, M., Rajesh, S., Girish, K., Johnson, Z., Anuradha, M., Gavaskar, S. S. M., Suneel, V. & Krishna, S. M. (2009). Reduced near-surface thermal inversions in 2005-06 in the southeastern Arabian Sea (Lakshadweep Sea). *Journal of Physical Oceanography, 39(5)*, 1184-1199.

Nodzu, M. I., Ogino, S. Y., Tachibana, Y. & Yamanaka, M. D. (2006). Climatological description of seasonal variations in lower-tropospheric temperature inversion layers over the Indochina Peninsula. *Journal of Climate, 19(13)*, 3307-3319.

Olofson, K. F. G., Andersson, P. U., Hallquist, M., Ljungström, E., Tang, L., Chen, D. & Pettersson, J. B. C. (2009). Urban aerosol evolution and particle formation during wintertime temperature inversions. *Atmospheric Environment, 43(2)*, 340-346.

Overland, J. E. & Guest, P. S. (1991). The Arctic snow and air temperature budget over sea ice during winter. *Journal of Geophysical Research, 96(C3)*, 4651-4662.

Overland, J. E. & Davidson, K. L. (1992). Geostrophic drag coefficients over sea ice. *Tellus Series A-B, 44A (1)*, 54-66.

Pérez, I. A., Sánchez, M. L., García, M. A. & de Torre, B. (2009). A classification of CO_2 concentrations based on a binary meteorological six variable system. *Agricultural and Forest Meteorology, 149(10)*, 1686-1692.

Phillpot, H. R. & Zillman, J. W. (1970). The surface temperature inversion over the Antarctic Continent. *Journal of Geophysical Research, 75(21)*, 4161-4169.

Preston-Whyte, R. A., Diab, R. D. & Tyson, P. D. (1976). Towards an inversion climatology of Southern-Africa: Part II. Non-Surface Inversions in the lower atmosphere. *South African geographical Journal, 59*, 45-59.

Prezerakos, N. G. (1998). Lower tropospheric structure and synoptic scale circulation patterns during prolonged temperature inversions over Athens, Greece. *Theoretical and Applied Climatology, 60*, 63-76.

Rigor, I. G., Colony, R. L. & Martin, S. (2000). Variations in surface air temperature observations in the Arctic, 1979-97. *Journal of Climate, 13(5)*, 896-914.

Russell, P. B. & Uthe, E. E. (1978). Regional patterns of mixing depth and stability: Sodar network measurements for input to air quality models. *Bulletin of the American Meteorological Society, 59(10)*, 1275-1287.

Serreze, M. C., Kahl, J. D. & Schnell, R. C. (1992). Low-level temperature inversions of the Eurasian Arctic and comparisons with Soviet drifting stations. *Journal of Climate, 5(6)*, 615-630.

Skony, S. M., Kahl, J. D. W. & Zaitseva, N. A. (1994). Differences between radiosonde and dropsonde temperature profiles over the Arctic Ocean. *Journal of Atmospheric & Oceanic Technology, 11(5)*, 1400-1408.

Soddell, J. R., McGuffie, K. & Holland, G. J. (2004). Intercomparison of atmospheric soundings from the aerosonde and radiosonde. *Journal of Applied Meteorology, 43(9)*, 1260-1269.

Sultan, S. A. R. & Elghribi, N. M. (1996). Temperature inversion in the Arabian Gulf and the Gulf of Oman. *Continental Shelf Research, 16(12)*, 1521-1544.

Timerev, A. A. & Egorova, S. A. (1991). Spatial-temporal variability of surface inversions in the Arctic. *Soviet Meteorology and Hydrology, 7*, 39-44.

Turner, D. B. (1970). Workbook of atmospheric dispersion estimates, U. S. Environ. *Prot. Agency, Research Triangle Park*, N. C.

Tyson, P. D., Preston-Whyte, R. A. & Diab, R. D. (1976). Towards an inversion climatology of Southern-Africa: Part I. Surface Inversions. *South African geographical Journal, 58*, 151-163.

Whiteman, C. D., Bian, X. & Zhong, S. (1999). Wintertime evolution of the temperature inversion in the Colorado Plateau Basin. *Journal of Applied Meteorology, 38(8)*, 1103-1117.

Yao, W. & Zhong, S. (2009). Nocturnal temperature inversions in a small, enclosed basin and their relationship to ambient atmospheric conditions. *Meteorology and Atmospheric Physics, 103(1-4)*, 195-210.

Zaitseva, N. A. (1993). Historical developments in radiosonde systems in the former Soviet Union. *Bulletin-American Meteorological Society, 74(10)*, 1893-1900.

Zaitseva, N. A., Skony, S. M. & Kahl, J. D. (1996). Temperature inversions over the western arctic from radiosonde data. *Russian Meteorology and Hydrology, 1(6)*, 6-17.

Zwoździak, A., Sówka, I., Gzella, A. & Zwoździak, J. (2008). Morning ozone buildup during summer 2003 episodes and their links with some radiosonde data in WrocŁaw, South-Western Poland. *Environment Protection Engineering, 34(1)*, 63-79.

In: Relative Humidity: Sensors, Management…
Editor: John M. Castillo
ISBN: 978-1-61761-734-8

Chapter 2

EVAPORATION AND EVAPOTRANSPIRATION, THEORY AND ASSESSMENT METHODS

S.H. Sanaeinejad
Ferdowsi University of Mashhad, Faculty of Agriculture, Mashhad, Iran

INTRODUCTION

Such as any other element, water also is evaporated. But since water plays a key role in our life on earth, its evaporation is very important in many aspects in different scientific fields. Water vapor is the main source of many atmospheric activities. It has an important role in the hydrologic cycle and water budget which is considered as the main solution of different water management problems. Evaporation has the core role in hydro-climatologic studies, watershed management, dams and water reservoirs designing. Evaporation is the most important parameter in irrigation programming and many other agricultural activities. It is important not only to prevent loosing water in drought but also to manage the consumption of water by plants, since it is the main source of loosing of water by plants through transpiration from stomata. Evaporation is most considered by agronomists in arid and semi-arid regions because limitation water availability is very restricted and evaporation is extremely high. It is also an important parameter in wet climates, when the need for investigation into its role in the water balance has been appreciated for a long time (Lemeur and Zhang, 1990).

For considering evaporation in any field of study we first need to know the basic physical theory of this phenomenon and also applying methods in different fields of studies.

PHYSICS OF EVAPORATION

Evaporation of water from a surface to form water vapour in the atmosphere is a complex physical process, involving both atmospheric and surface parameters. For instance the latent heat of water vaporisation is large in relation to the specific heat of air, soil and most other materials. This characteristic is due to the hydrogenous link between molecules of water. For this reason water can save energy during the day and release it, part of that via evaporation, during the night. This process causes small range in temperature between day and night in the areas close to sea.

Although research on evaporation into the atmosphere has been carried on since the ancient ages, perhaps it could be said that a formulized description of evaporation was first studied by Dalton about 200 years ago. He formulized evaporation in respect of vapour pressure deficit. He also found that the evaporation from a water surface is directly proportional to the vapour pressure deficit, under especial conditions when the other factors remain constant (Brutsaert, 1982). Later, Rowher (1931), Brunt (1939), Sutton (1934) and Pasquill (1943) used the Dalton equation to state the effects of atmospheric parameters on evaporation. Penman (1948) used this equation with own formulation for the energy balance to find a combination formula which is still the most complete physical description of the evaporation process.

The concept of evaporation is in general use and when used for special purposes it must have exact definition. Although, a lot of work has been done in this field, the different conceptions and terminology of the area are not standard and it is necessary to define them in every study to prevent confusion. Some of these concepts which express a considerable part of the history of this field are as follows:

Evaporation: Evaporation is the physical process by which water is transferred from its liquid phase to gas (Jensen et al., 1990).

Potential evaporation: When all the evaporating surfaces are wet, so there is no restriction on water in the evaporation process from the surface, it is called potential evaporation (Jensen et al., 1990)

Nevertheless, there are some modifications to this conception. Granger (1989) presented five different conceptions of potential evaporation. He showed that "not only are there several different definitions in use for identifying the parameter called potential evaporation, but also that these can represent different, distinct and definable terms. This in fact represents the major source of confusion surrounding the concept of potential evaporation". The Penman potential evaporation is in the concern of this study is defined exactly to prevent this sort of confusions.

Evaporation depends only on atmospheric conditions, When there is no water deficit and evaporation is in the potential rate.

Atmospheric Factors

The atmospheric factors affecting evaporation can be referred to the energy requirement to convert water into vapour and the mechanism which determines the transfer of vapour into the atmosphere (Siam, 1979; and Doorenbos and Pruitt, 1984). Hence, these two factors are described.

The Energy Requirement for Evaporation

Evaporation is a process which takes energy to convert water molecules from liquid phase to gas phase. Through this process 540 calorie energy is necessary to convert one gr. water from liquid phase to gas phase in 100 °C (600 Calorie/gr. in 0 °C) (Monteith, 1990). Consequently, if the amount of energy, which is consumed for evaporation, is known the amount of evaporation will be calculable. This simple relationship between available energy and evaporation has been one of the important branches in evaporation studies. The energy balance equation can be written as follows (Penman, 1948):

$$R_n = \lambda E + H + G \quad (1)$$

In which R_n is the net radiation, at the surface, λE is the latent heat flux, H is the sensible heat flux to the air and G is the heat flux to the soil.

The major problem in the energy balance equation is determination of the net radiation at the surface. The importance of accurate knowledge of surface radiation budget is well known and has been discussed widely. Although, there is a long history of ground-based radiation measurements, there are many areas with lack measurement and others where the spatial coverage is, at best, sparse and for large regions entirely absent. There are, for example, few measurements over the two-thirds of the earth's surface that is covered by oceans, as well as over the Polar Regions (Zhanqing et al., 1993). However most of the times, it is required to have a detailed spatial distribution of radiation over a surface with a variety of slope and aspect, where there are few point measurements. Therefore some authors have attempted to obtain this by estimation equations which include other available weather parameters such as sunshine hours, percentage of cloud etc. (MTSAT-1R, 2010)

However, the point measured radiation data, which has been traditionally used in climatology studies, is still used for some site studies applications. Years ago, an empirical equation was introduced by Prescott (1940) to estimate incoming short-wave radiation as follows:

$$R/R_0 = a + b\,(n/N) \quad (2)$$

Where R is the global short-wave radiation (J/cm^2), R_0 is the solar radiation (J/cm^2), on a horizontal surface under clear skies condition, n is the sunshine hours and N is daylight hours; a and b are constant. This equation has been used by other workers and the coefficients a and b has been determined for different areas. Black et al., (1954) established a relationship which has considered the effect of latitude on the coefficients in the Penman equation. Hanna and Salem (1980) used the daily sunshine hours and measured daily global radiation, for a decade period and showed that the a and b coefficients are 0.26 and 0.46 respectively. This differs from the coefficients which were calculated 0.17 and 0.52 for a and b respectively using the equation of Black et al., (1954).

Reflection by the surface, the radiation of the surface corresponding to the Plank radiation (Monteith 1990) and even the amount of aerosols in the atmosphere also influence at the Earth's surface radiation. Stringer (1972) recognised that absorption of radiation is negligible in summer and is about 5 to 6% in winter in some country districts, but reaches 30 % or more in industrial areas, depending on smoke concentration in the atmosphere.

There are several factors affecting radiation and consequently energy balance and evaporation, some of the most important are as following

1. Short-wave incoming radiation,
2. Reflection of short-wave from the Earth surface,
3. Diffusion of short-wave by atmosphere,
4. The fraction of radiation used in photosynthesis,
5. Long wave emission by the Earth surface and
6. Cloudiness.

There are other factors influencing radiation such as: sun spots, climate change etc. which are not in the concern of this chapter. The diffuse sky radiation varies greatly even within a small area and although it is often neglected in the study of the radiation balance; its effect can be approximated by pyrometers which measure scattered and global radiation (Stringer, 1972). Although the diffused radiation is negligible in a sunny day, it can be the major part of total radiation in a cloudy condition; therefore it has to be considered as a part of total radiation (Sanaei-Nejad and Hanna, 1995).

Long wave emission by the surface is another component of radiation which has been discussed by some workers in their studies. This term can be calculated by the Stephan-Bultsman Equation (Penman, 1948 and Monteith, 1965) as follows:

$$R_L=\delta T^4 f(d) \tag{3}$$

In which R_L is the long wave radiation, δ is the Stephan-Bultsman constant, T is the temperature in °K and $f(d)$ is used to take into account the effect of cloud. For the surfaces within a small area and with approximately equal reflection coefficient, mean monthly total radiation depends on local amount of cloudiness (Stringer, 1972).

Temperature is another atmospheric parameter which has a secondary role in the energy balance equation. If there is no advection, the amount of sensible heat is determined by measuring air temperature. Penman (1948) and Monteith (1965) used this parameter to calculate long wave radiation and considered its effect on vapour pressure deficit. Some other authors used temperature as an independent factor for their evaporation studies (Thornthwaite, 1948; and Blaney and Criddle, 1950). However such experimental equations have proved unreliable to be applied for other areas.

Another atmospheric characteristic, which affects evaporation through the energy balance equation, is advection of sensible heat. Advection causes to transfer sensible heat from a region to another region and consequently changes the energy balance. The "oasis" effect and "clothesline" effects are two important results of this procedure. The earliest occurs when sensible heat transfers from a dry zone to an irrigation zone and the latter is the same occurrence on vegetation cover which causes anomalous evaporation rates (Barry, 1973).

There is other complication in evaporation due to temperature. This is when minimum temperature is less than dew point. In this case the condensation take place, and some energy will be released and when the temperature rise again the dew will evaporate before the mainly evaporation.

The Mechanism of Transferring Vapor to the Atmosphere

Dalton's equation (Dalton, 1802) was the first formulation which was concerned with the aerodynamic effects in the evaporation process. He considered the gradient of vapour pressure above evaporating surface and horizontal motion of air mentioned as a function of wind speed. The equation is as follows (Penman, 1948):

$$E = (e_s - e_d)\, f(u) \tag{4}$$

In which e_s is vapour pressure at the surface, *ed* is the vapour pressure in the atmosphere above the evaporating surface and *f(u)* is a function of the horizontal wind velocity.

Penman used this equation and made some changes on it so that the included parameters were more measurable. He assumed that transport of vapour and the transport of heat by eddy diffusion are, controlled by the same mechanism. For this reason he used the Bowen ratio in the form as follows:

$$\beta = K/E = \gamma\, (T_s - T_a) / (e_s - e_a) \tag{5}$$

where K is the sensible heat, E is the evaporation rate, γ is the standard constant which is known as psychrometric parameter, T_s is the surface temperature T_a is the air temperature, e_s is the vapour pressure at the surface and e_a is the vapour pressure in the atmosphere.

The aerodynamic effect depends on molecular diffusion of vapour into air. The motion of air could be horizontal as created by wind speed, and vertical as under convection conditions. Penman (1948) was concern about process of molecular diffusion which takes place in a non-turbulent thin layer of air, but he did not consider this in his formula to retain its simplicity and make it more easily applicable. He used f(u), a function of wind run at 2m high (u_2) and recognised that the best form of the function for practical use is as follows:

$$f(u) = 0.35\, (1 + 9.8 \times 10^{-3} u_2) \tag{6}$$

The free convection regime can be expressed as a function of the canopy-air temperature (Zhang and Lemeure, 1992).

Evapotranspiration: Evaporation occurring through the plants water use called transpiration. Although transpiration is a type of evaporation because it is a process through which water transfers to vapour; it should be considered as a separate process, because of the effect of the plants physiology on the evaporation process. For this reason "the combined process by which water is transferred from the earth's surface to the atmosphere; evaporation of liquid water from the soil surface and water intercepted by plants, plus transpiration by plants" is called *evapotranspiration* (Jensen et al., 1990).

Potential Evapotranspiration

This is the rate at which water would be transferred from wet soil and plant surface to the atmosphere when there is no restriction of water. The concept of potential evapotranspiration was introduced by Thornthwaite (1948) in the context of the classification of climate. He based this in the assumption that "there is a distinction, then, between the amount of water that actually transpires and evaporates and that which would transpire and evaporate if it were available, when water supply increases, as in a desert irrigation project, evaporation rises to a maximum that depends only on the climate. This we may call '*potential evapotranspiration*', distinct from actual evapotranspiration".

Based upon the FAO (Doorenbos and Pruitt, 1984) definition potential evapotranspiration can be considered as "the theoretical evapotranspiration which would occur from an infinite expanse of natural or cultivated terrain, adequately supplied with moisture exposed to specified climatic conditions which remain unaltered by the evapotranspiration itself."

The main factor in definition of potential evapotranspiration is the plant types and conditions, which are discussed here.

Plant Factors

Plant factor has been considered in the earliest study of evaporation. Penman (1948) in his experimental work at Rothamsted used some cylinders that could be called a type of lysimeter, two of which included turf and he studied the effect of this type of vegetation on potential evaporation. The rate of potential evaporation from grass surface is different from water surface and the ratio varies during the year and other changes in environmental conditions. Penman (1948) showed that the ratio of evaporation from grass surface to the evaporation from water surface varies from 0.7 in summer to about 0.3 in winter.

Monteith (1965) modified Penman equation and considered a coefficient, rc, in which he included stomata resistance of leaves against the transpiration process. Wright and Harding (1993) recognised where the grass is completely active, during July-September, the water use is very close to Penman evaporation, E_t. However, in the early part of summer, during April and May, the evaporation is reduced to between 50-60 % of E_t, as a result of the dormancy of the grass.

Some other workers such as Dickinson et al., (1991) have distinguished between soil and canopy temperatures. Zhang and Lemeur (1992) also stated that the characteristic dimension of the leaves affect on the aerodynamic resistance which is an effective atmospheric parameters in evaporation.

Reflection of solar radiation from plants is also different depending on the type and the density of the plants. This term is defined as "*albedo*" and has to be derived for any different plant cover. Difference in the surface albedo causes a different energy balance which greatly affects evaporation Actual evapotranspiration.

The conditions for potential evapotranspiration are not available at most of the times and most of the areas. Therefore the amount of actual evapotranspiration is not always equal to potential evapotranspiration. The rate of actual evapotranspiration is a function of the vegetation canopy and soil moisture. Stomata closure due to temperature, water or even nutritional stress can sometimes reduce actual evapotranspiration by a factor of 10 times below potential evapotranspiration.

Determination of actual evapotranspiration is one of the most problematic subjects. However, potential evapotranspiration has become a standard ingredient for estimating actual evapotranspiration (See: Granger, 1989; Green, 1970; Calder, 1990).

For estimating actual evapotranspiration it is necessary to make a preliminary survey of how much lower actual evapotranspiration is than potential evapotranspiration throughout a special region. For example Mondeville (1990) found a range from a minimum of 39 % to a maximum of 67 % for an area in the East Africa.

Several reports have been published in the subject of estimation of actual evaporation consequently there are numerous experimental equations available (See: Thornthwaite, 1948; Blaney and Criddle, 1950; Makkink, 1975; Brutsaert and Stricker, 1979). Monteith (1965) suggested an approach to modelling actual evaporation that included a measure of the resistance to the diffusion of water vapour by air and crops. (See also Granger, 1989; Mandeville and Batchelor 1990; Calder, 1990 and Paralange and Katul, 1992).

One of the factors affecting actual evaporation is horizontal transfer of energy called advection. Paralange and Katul (1992) have discussed the effect of local advection on actual evaporation and gave a methodology to account for the excess advection in the complementary model. Soil moisture and its conditions are other important issue to be considered in the actual evaporation.

Soil Factor

For bare soil loses of water by evaporation is very low and it may be negligible when the thin layer of the surface has dried out. Nevertheless, the soil water content has an important role in actual evaporation process through the plants. Water evaporates directly from the wet soil surface and extracts by transpiration through the roots of plants (Denmead and Show, 1962; Veihmeyer, 1927; Van Bavel, 1953). Both of these processes as depend on physical and hydrodynamic characteristics of the soil. The actual evaporation is divided into two stages. The first stage lasts until cumulative evaporation reaches a given threshold, dependence on the soil texture; Ritchie (1972), has given some values for this which are 6mm for sandy soils, 9mm for loamy soils and 12mm for clayey soils. During this stage, evaporation depends on climatic demand only. During the second stage, evaporation from soil is reduced with respect to the potential reference. The extent of this reduction depends on water content of the surface layer as well as on the hydrodynamic properties of the soil (Brisson and Seguin, 1992).

Interception evaporation: When there is a rainfall over a canopy, evaporation take place from surface leaves, branches and trunks (in forest) as well as evaporation that take place from soil and transpiration; this is called *interception evaporation (*Shuttleworth, 1983). It can be a considerable portion of total evaporation depends on type of vegetation in the study area. It is obvious that the amount of interception evaporation is more in forest than shorter plants. (See Rutter et al., 1971; Gash and Morton, 1978 and Calder, 1976). It is dependent on the frequency and the amount of rainfall. Therefore it can be one of the most important parameters in the study of rainfall regime in a forested area (Gash and Stewart, 1977). Rutter (1963, 1967) and Stewart (1977) showed that evaporation from a forested area will be at rates in excess of the potential evaporation rate from wet grassland. Calder (1976) used a natural lysimeter in central Wales and showed that only 310 mm of 1110 mm water loss was made up from transpiration and the rest was because of the interception process. He concluded that "the annual loss recorded at the lysimeter, amounting to twice Penman E_0, demonstrates the magnitude of the evaporation losses that can arise from forests in a high rainfall area". As the major part of the interception evaporation occurs from forests than short vegetation, Calder (1976) also concluded that the larger forested area in a watershed the higher interception loss of rainfall would be. (See also Gash and Stewart, 1977 and Gash and Morton, 1978). Then he introduced the following exponential equation to estimate interception evaporation from heather and forest in the mountainous area of Balquhidder in Scotland (Calder, 1986).

$$I = \gamma(1 - e^{-\delta P}) \quad (7)$$

Where γ is the maximum predicted interception loss in one day, σ is an experimental constant and P is the daily precipitation.

The range of interception evaporation over Great Britain is of the order of 600mm, while potential evaporation range is of the order of 300mm. "This difference is primarily due to the different duration of rainfall, and hence of canopy wetness in the two area" with different interception evaporation (Gash and Morton, 1978). Nevertheless, there are wide ranging values of interception in different climatic conditions reported in the literature (Persson and Lindroth, 1994). For example, 10 to 51 % of rainfall has been reported for coniferous forests (Helvey, 1971, Mulder, 1983) while this is 9% for tropical rain forest (Lloyd et al., 1988) and 11-18% has been derived for acacia (Bruijnzeel and Wiersum, 1987).

So, interception is a critical hydrologic factor, because it determines the amount of water remains in the system after precipitation continuing to storage in the soil or to stream flow water (Running, 1991). (See also Gash, 1980; Calder, 1990 and Johnson, 1990).

The Nature of the Evaporating Surface

The type of the evaporating surface is also affect evaporation. Penman (1948) used three different types of evaporating surface (free water, bare soil and crop covered surface) in his experimental work at Rothamsted. There are more evaporating surfaces in nature, water may evaporate from a surface which is covered with snow, rough and smooth surfaces also produce different evaporation conditions, it also differs according to vegetation types, short plants, like grass and alfalfa, and long vegetation, like trees and forests.

The first characteristic which causes a difference in evaporation from different types of evaporating surface is *albedo*. Albedo is defined as a percentage of reflected short wave radiation by surface. The regional surface albedo varies with ground cover, soil type, crop type and crop stage of growth (Shuttleworth, 1986 and Rowntree, 1988). Penman (1948) considered the albedo as six percent for water, one percent for bare soil and two percent for turf covered surface in the British Isles. These values are annual averages but noted that they would vary with season.

The difference albedo depends on different crop types and the elevation of sun (Stringer, 1972). Doorenbos and Pruitt (1984) reported a range between 5-8 % for water reflection and 20-25 % for vegetation surface, dependence on the factors. (See also: Mondeville, and Batchelor, 1990). Calder (1990) used the amount of 13 % for heather and 25 % for grass.

The albedo for snow covered surfaces is more than other surfaces. The result of the much higher albedo of snow is that the amount of energy, which is available for evaporation from snow covered surface, will be less than the energy available for wet vegetated surface under the same external meteorological conditions (Hall, 1987; Calder, 1990). Hall (1987) also showed that the value of evaporation from snow covered parts of a watershed in the UK could be assumed zero. Total evaporation over a snow covered area in a three weeks period has been recorded as less than 1mm (Hall, 1987).

Differences in types of evaporating surface also affect terrestrial radiation. Stringer (1972) has explained that local radiation emission from short turf is close to black body radiation at the temperature of grass-minimum temperature. Another source of difference in the energy balance equation is the absorbency of energy by the surface. Since the soil heat term is not considered in the Penman equation, it might affect the accuracy of the equation, because this term is not always negligible, particularly, when short term energy balance is considered (Jaywardena, 1989).

Different types of evaporating surface have different aerodynamic effects on the evaporation process. For example the turbulence in rough areas is more than in smooth areas. It is well known that the wind speed increases with altitude, then its effect on evaporation rate would be increased in mountains (Wright and Harding, 1993). This effect also occurs because of difference in height and roughness of vegetation. Evaporation from short vegetation is controlled by meteorological parameters above the surface, unless there is a water stress in the plants. So, the Penman equation can be used to estimate the water loss from the short plants. However, the aerodynamic roughness of forest generates an intensive turbulence increasing the vapour flux from the forest's leaves (Gash, 1977).

Sudanese Papyrus may evapotranspirate twice as much as free water surface. In some cases evapotranspiration values are even 3 to 4 times higher than potential evaporation from free water surface which has taken place by special swamp vegetation Balek (1977)

Another characteristic of different surfaces important to the evaporation process is water availability. It is obvious that there is no restriction in evaporation on a free water surface. However, evaporation from most bare soils, in reality, is greatly reduced after the loss from the surface layer of about 1cm of water (Dickinson et al., 1991). The conditions are completely different for a plant; its root system can draw on moisture throughout a deep layer of soil and its aerial parts cause it to transfer vapour throughout a considerable thickness of air, though its photo-sensitive stomata mechanism can be a limiting factor in the evaporation process (Penman, 1948; Calder, 1990; Dickinson et al., 1991). Therefore, the amount of evaporation from vegetation is different from bare soil surfaces and free water surfaces. Also evaporation differs between types of plants and even during plant growing stages in the same plant (Brisson and Seguin., 1992). The plant coefficient is a function of relative humidity, wind speed and harvesting interval and it should be derived for each plant type separately (Doorenbos and Pruitt, 1984 and Jensen, 1973).

APPLICATIONS OF EVAPORATION STUDIES

Evaporation studies are applicable to a wide range of practitioners. Some of the most important of the applications are for agriculturists, foresters, hydrologists, climatologists, meteorologists and plant physiologists. For instance in agricultural studies the rate of evaporation influences the time of seeding, scheduling of irrigation, herbicide and pesticide applications and tillage practices. It is an important parameter in climate classifications and also its local effect in climate over time, distance, altitude, advection, soil water availability. Some evidences have been also accumulating from climate model experiments that changes in the evaporation at the land surface affects climatic change. (Rowntree, 1991).

Some examples in application of evaporation in hydrological studies are:

- Estimation of reservoir evaporation,
- Evapotranspiration as a component of the water balance,
- In arid zones, estimation of evaporation from shallow depressions following heavy rainfall, and
- Some other subjects include planing of channels when the range of water losses from natural watersheds is important. (Jones, 1981 and Calder, 1990)

Spatial Scale in Evaporation Studies

Much work has been carried out in different disciplines in the extending the application of evaporation studies. Nevertheless, the spatial scales of these studies have been different. The scales of evaporation studies have been ranged from global to continental, regional, local and micro scales. In the global scale, the studies have been concerning with the role of evaporation in the hydrologic cycle and changing global climate. It is obvious that decreasing evapotranspiration rate causes a decreasing rate of rainfall in a pattern of global hydrologic cycle (Balek, 1977, Schmugge and Andre,1991).

For many applications, the regional scale studies are more useful. Though, for some others small-scale studies such as a watershed or a suburban area are more applicable. However, the problem is that the results from watershed scale experiments cannot be applied elsewhere. These watershed scale results are limited in value for making a suitable assessment of the effects of a change of land use on evaporation in a limited area with similar climatic conditions (Calder, 1990).

Time Scale in Evaporation Studies

The time scale of evaporation studies also differ widely. Some research has been done by researchers in the scale of less than an hour and some others in the period of more than ten years. For example Hellwig (1973) made an experiment to show that near Windhook, evaporation from moist sand was highest just before sunrise (the time scale smaller than an hour). Suckling (1980) selected a daily range to study about suburban evaporation and his experiment was performed during a few significant days in his chosen area. Malek (1992) studied the night-time evapotranspiration versus day time and 24 hours evapotranspiration of alfalfa in northern Utah (US). His measurement time intervals were 20 minutes during a few days of a complete growing cycle of alfalfa in August and September. Parlange and Katul (1992) also measured potential evaporation from a wet bare soil field on a 20 minutes time step for 5 days at Davis, California (US). Jones (1981) analysed the values of evaporation taken by a regression equation for 72 sets of 10 days totals. He also studied the seasonal amount of potential evaporation and rainfall and recognised their variation with latitude in low latitudes. Penman (1948) established his equation based on two years measurements which had been done with one day intervals. He also considered to some seasonal changes. Wright and Harding (1993) made an assessment for seasonal variation of water use of grass and Penman evaporation in the highlands of Scotland. He concluded that in the early part of the summer evaporation is reduced to between 50-60 % of Penman evaporation and in the early month of the year is strongly related to the vitality of grasses. Doorenbos and Pruitt (1984) recommended that calculation of mean daily evapotranspiration by Blaney-Criddle should be made for periods no shorter than one month. Calder (1986) and Giambelliuca and Nullet (1992) discussed and recognised some characteristics for seasonal variation of evaporation.

On the other hand, some other workers like Smith (1967) used a 15 years period of evaporation data to study evaporation in UK and Mondeville and Batchelor (1990) used a 10 years period of evaporation data to study estimation of actual evapotranspiration in Malawi (Africa). However, FAO preferred to calculate evaporation data for each month or period for each year of record rather than using the mean value data on several years of records. Because of changing climatic conditions for each month or shorter periods from year to year, except for equatorial zones (Doorenbos and Pruitt, 1984).

Evaluation of Evaporation

The data, which is needed for evaporation studies, usually has been obtained by three methods.

- Measuring of evaporation directly by some equipments
- Estimation of evaporation by modelling and statistical regression equations
- Using Remote Sensing data to evaluate evaporation

Measurement of Evaporation

Many devices have been proposed and developed by different workers over the past few centuries to measure evaporation. There have been several pans and wet surface devices, but most of them are no longer in use today. *The class A Pan of the US Weather Bureau is the most used device which is currently used in all of the countries.* WMO introduced American class “A” pan evaporation as a standard means for measurement evaporation (WMO Tech note No. 83). "The class “A” evaporation pan is circular, 121cm (46.4 inches)in diameter and

25.5cm (10 inches) deep. It is made of galvanized iron (22 gauge) or monel metal (0.8mm). The soil is built up to within 5cm of the bottom of the pan. The pan water level should not drop to more than 7.5cm below the rim. The pan if galvanized is painted annually with aluminum paint" (Doorenbos and Pruitt, 1984).

There are other types of evaporation measurement devices beside pans, such as evaporimeters and atmometer. The most common of them are the Pich evaporimeter, the Wild evaporimeter and the porous cup atmometer. Since the evaporation data derived by using these equipments are not as accurate as pans, they are not widely used. More details about these equipments can be found in any meteorological equipment manuals.

Measurement of Evapotranspiration

Pan evaporation is physically different from evapotranspiration from vegetation and soil surfaces. Nevertheless, in some special conditions, when the evaporating surface is not exposed to heat advection, there is a high correlation between pan data and potential evapotranspiration from its surrounding area. Thus under this special condition it is possible to use pan evaporation data to estimate evapotranspiration (Penman, 1948; Brutseart, 1982; McLlory and Angus, 1964). Otherwise, it is necessary to measure evapotranspiration directly by some special equipment, since the pan evaporation data cannot be used when advection occurs and also when the pan coefficient is not known. In this section some of these equipments are explained.

Neutron probe method: Determination of actual evapotranspiration under field capacity is possible by measuring the change in soil water over a period of time (Jensen et al., 1990). This can be achieved by using neutron soil water probe equipment. The neutron probe technique is one of the most accurate soil water content measurement methods (Osmar, and Cuenca, 1992). It works based on a theory that water molecules cause a reduction in the speed of neutrons. There is an atomic element inside the equipment to release high speed neutrons and a detector to count the low speed neutrons reflected from water molecules inside the soil body. This is complicated equipment and care must be taken to calibrate and use it properly. Although the water content of soils can be evaluated accurately by this equipment, there are some restrictions to use it to determine the soil moisture in short period of time such as daily measurement.

1. ***Lysimeters:*** Lysimeter is a container filled by soil and planted and placed in a field with the same vegetation grown under natural conditions. The evapotranspiration can be measured using this tank by different methods. In a general classification "Lysimeters can be grouped into three categories: (1) non-weighing, constant water-table type which provides reliable data in areas where a high water table normally exists and where the water table level is maintained essentially at the same level inside as outside the Lysimeter; (2) non-weighting, percolation type, in which changes in water stored in the soil are determined by sampling or neutron methods and the rainfall and percolate are measured (often used in areas of high precipitation); and (3) weighing types in which changes in soil water are determined

either by weighting the entire unit with a mechanical scale, counterbalanced scale and load cell, or by supporting the Lysimeter hydraulically" (Jensen et al., 1990).

2. As the applications of Lysimeters are different and depends on the special circumstances and the type of studies, there is no definition for a standard Lysimeter. However, care must be taken to select a Lysimeter so that it can represent the evapotranspiration of its surrounding area. For this reason the size and the type of Lysimeters are widely different so that some workers have been using a mini Lysimeter which is circular and constructed of sheet metal, 0.25m in diameter and 0.15m in depth while there are some other Lysimeter being used with more than 6 meters in diameter and 3.5 meters in depth. Calder (1976) used a natural drainage Lysimeter with 84 m2 plot and 10m height to measure evaporation from a forested catchment.

Estimation of Evaporation

Evaluation of evaporation by measurement is not always possible everywhere because of several reasons some of which are as follows:

- Sparse network: There are not enough measurement stations in most regions to produce a satisfactory network.
- Measurement of evaporation from large area is not possible by any equipment. The most satisfactory would be a large lake or vegetated area but it is only possible to measure from a small controlled site.
- There are some systematic errors in measuring of evaporation that could not be negligible. Some of which appear in measurement by the Pan evaporation are:
- Pan water takes more energy than free water surface because of conducting sensible heat through its body.
- The temperature of water in the Pan increases more than a large body of water.
- Heat transfer occurs through the sides of the Pan.
- The color of the Pan affects radiation absorbency and consequently the amount of evaporation.
- Advection affects the amount of evaporation dependence on the situation of the Pan.

For measuring evapotranspiration from vegetation area a Lysimeter can help, but because of the high costs of a Lysimetric station, these data is very rare. Therefore, it is necessary to estimate evaporation by special methods. Using these methods, the amount of evaporation can be derived in respect of other meteorological factors. One of the earliest studies in this regard back to 1809, when Dalton recognised the correlation between vapour pressure deficit, wind speed and evaporation. Then in 1913 Bridges and Shantz recognised the effects of radiation and sensible heat on this process (Jensen et al., 1990). Penman (1948) completed this study using the conception of energy balance for evaporation from a free water surface, which had been developed in the 1920s, on a cultivated land.

At the present there are several models and equations to estimate evaporation and evapotranspiration. However, they all can be divided into two major classifications. The equations which have been established by theoretical conclusion and those are empirical and have been obtained by experimental data analysis.

Estimation of Evaporation by Modeling Based on Physics of Evaporation

It is possible to divide all the theoretical methods into five general categories:

1. The water balance or hydrological method.
2. The eddy correlation method.
3. The energy balance method.
4. The aerodynamic or profile method.
5. The combination method.

Although these methods have a theoretical basis, all require experimentally derived elements in order to quantify the relationship.

The Water Balance Method

The base of this method is the conservation of mass applied to some parts of the hydrological cycle. This can be formulated, in general, for any area by the following equation:

$$E = P - Q - G \tag{8}$$

In which E is the evaporation, P is the precipitation, Q is the runoff and G is the percolation water in the ground. Using the equation, evaporation can be estimated, if all of the other terms cannot be determined independently. One of the major problems in using this method is that it can be used to estimate evaporation in short term period (Jensen et al., 1990). This method is generally used on large areas such as watersheds, where the outflow can be determined from measurements in the discharge point of the watershed and inflow is measured using rain-gage network stations.

The Eddy Correlation Method

Water vapor process can be formulated using fluid dynamic theory. This theory considers the water transfer into the atmosphere under two motions, first through convection with the air turbulence and second, through molecular motion. The original equations resulting from this theory are very sophisticated and not easy to solve. Simplifying these equations a more practical equation can be derived whic is as follows:

$$E = \rho w' q' \tag{9}$$

in which E is water vapour flux, (is the density of the air in the surface, w' is the vertical turbulence and q' is the specific humidity (Brutsaert, 1982).

Some special equipment are necessary to measure w' and q' in the site based on we could calculate evaporation from this equation. There are "many methods which have been applied to measure these fluctuations. For the vertical fluctuations hot wire-type anemometers, which are probably used first, are available commercially. Other instruments for this purpose are sonic anemometers" (Brutsaert, 1982).

The major problem in this method is complexity in the measurement of the necessary data.

The Energy Balance Method

The basis of this study is to determine the energy consumed by the water to be evaporated. The principal source of this energy is solar radiation. However, it is the net radiation which is the climatic factor controlling evaporation under potential conditions when there is no water limitation for evaporation. Therefore, estimation of net radiation is the basic stage of the stimulation of evaporation by this method. Numerous studies have been published estimating evaporation using the energy balance equation (1) (Penman, 1948; Priestley and Taylor, 1972; Hanna and Siam, 1980; Shuttleworth and Wallance, 1985; Lafleur, 1992 and many others).

Net radiation can be measured or determined using other radiation components data and the following equation.

$$R_n = (1 - \alpha)R_s - R_L \tag{10}$$

Where α is the albedo, R_s is the solar radiation and R_L is the terrestrial radiation into space or can be estimated from some experimental equations as earlier was explained.

If R_n is known the sensible heat, H, can be determined by some independent methods. One of which is a combination of the energy balance equation (1) with the Bowen ratio producing the following equation:

$$E = R_n / (1 + \beta) \tag{11}$$

In which R_n is the net radiation, β is the Bowen ratio. Determination of β requires measurement of temperature and specific humidity at two levels in the air applied in equation (4). When these data are not available it is possible to use the mean wind speed profile in the application of the energy budget method (Brutseart, 1982).

This method can be used for determination of evaporation for hourly or shorter periods, especially during daylight hours. However, the requirement equipment for measuring the necessary data is the main reason for restricting the use of the energy balance method in the estimation of evaporation (Jensen et al., 1992).

The Aerodynamic or Profile Method

As the water vapour is one of the components of the lower atmosphere, the dynamic of sub-layer of the atmosphere can describe the evaporation procedure properly. In the atmosphere most of the changes in atmospheric parameters such as temperature, humidity and wind occur in the vertical and very close to the surface (Brutsaert, 1982).

Although, the atmospheric motion in the sub-layer is very complicated, the atmospheric boundary layer (ABL) theory simplifies these motions to formulate the water vapour and the other atmospheric parameters changing in vertical profile. The measurement techniques needed in this method make it practically difficult.

The Combination Method

The original Penman equation is the oldest theoretical equation which considers all the different parameters affecting evaporation. He considered physics of evaporation, the available energy and the aerodynamic characteristics to give a combination method based on a relatively perfect discussion of evaporation theory. The equation is as follows (Penman, 1948):

$$E= (\Delta R_n + Ea\gamma)/ (\Delta+\gamma) \tag{12}$$

Where E is the evaporation in unit time, R_n is the net radiation energy available at surface, Δ is defined by:

$$\Delta=de_a/dT_a \tag{13}$$

In which e_a is the saturation vapour pressure at the air temperature (T_a), γ is the psychrometric coefficient ($\approx$ 0.27) and E_a is defined by:

$$E_a= (e_a-e_d)f(u) \tag{14}$$

In which e_d is the vapour pressure at the dew point and $f(u)$ is a function of wind speed.

Penman during his experiment also concerned very carefully the surrounding conditions. He considered plant cover, type of plants, the nearest distance of obstacle against the wind, such as trees, wind direction and even the stage of growing plants effects on the evaporation process. He also pointed out the discontinuity which the enclosure had caused, but he did not remove this and neglected the effect during his experiment (Penman, 1948).

The Penman equation is the simplest of estimation of evaporation which has been introduced based on physical theory (Shuttleworth, 1983). It was chosen as the reference potential evaporation "because not only are values readily available but because it also gives reasonable estimates of grass evaporation (for which it was calibrated) and of transpiration from spruce forest." Calder (1990).

FAO also recommended that the Penman method provides the evaporation estimation for areas where measured data on temperature, humidity, wind and sunshine duration or radiation are available. The Penman method offers the best results with minimum possible error which is not more than 10 % in summer, and 20 % under low evaporative conditions (Doorenbos and Pruitt, 1984).

Some workers have studied the Penman method and made some modifications to improve it. Van-Bavel (1966) suggested the same form of the equation of Penman but he changed the aerodynamic term of the equation into this form:

$$E_a=315u\ (e_s-e_d)/T[ln\ (z_a-z_o)]^2 \tag{15}$$

Where T is the air temperature (°k), z_o is the roughness length of the surface, and z_a is the wind measurement height (Giambelluca and Nullet, 1992). Priestly and Taylor (1972) neglected the aerodynamic factor in the Penman equation and introduced the energy term of the equation as an independent estimation equation.

One of the most important works base on the Penman method has been performed by Monteith (1965). He suggested a new approach to model E_a that included a measure of the resistance to the diffusion of water vapour. He added some respected parameters into Penman's equation and wrote his new equation as follows (Monteith, 1990):

$$\lambda E=[\Delta R_n+\rho C_p\ (e_s-e)/r_a]\ /\ (\Delta+\lambda\ (1+\ r_s\ /\ r_a) \tag{16}$$

Where λE is the evaporation expressed as latent heat flux (Wm-2), C_p is the specific heat of air at constant pressure (J kg^{-1} k^{-1}) ρ is the air density (kg m^{-3}) and r_a is the aerodynamic resistance to water vapour diffusion (sm^{-1}) and r_s is the surface resistance. The other notations are the same as equation (12) Monteith (1990) explained that the equation is valid for any surface at which the free energy of water can be treated as a constant.

However, he did not recommend this equation for estimating evaporation from bare soil, because water content changes very rapidly with depth below the surface. This equation has been called as Penman-Monteith equation, because the main part of the Penman equation has remained and Monteith has made some useful changes in the equation.

Coefficients, ra and rs are the most prominent characteristics of the equation, so that they determine the amount of aerodynamic parameters in the equation. Calder (1990) explained that if this equation is considered to estimate typical forest evaporation inserting ra and rs in the equation, it appears that the transpiration will be dominated by the humidity deficit of the atmosphere rather than the net radiation. Whereas, the same consideration with the Penman equation will show that the radiation term is usually dominant. For the reason the Penman equation is more useful to estimate short plants evapotranspiration than long vegetation (Calder, 1990). On the other hand, it can be a very useful method to calculate evaporation in a watershed with different vegetation cover.

Estimation of Evaporation Based on Statistical Regression Models

The theoretical based models usually require several environmental parameters. This characteristic causes some difficulties with estimating of evaporation by these methods some of which are as following:

1. All of the required data are not always available,
2. For some applications simple equations are more appropriate than the complex theoretical equations.

Therefore, in some cases within the range of applicability of the empirical (or semi-empirical) methods "somewhat paradoxically, the semi-empirical methods may actually provide more accurate estimates of long term evaporation losses than the more theoretical methods. This is because of deficiencies in the instrumentation used for measuring the meteorological variable" (Calder, 1990).

For these reasons, some workers have tried to establish some empirical estimation of evaporation equations which require less data and are simpler than theoretical equations.

However, factors such as availability of the required data for estimation of evaporation must be considered when choosing the evaporation calculation technique (Shih et al., 1983 and 1991). These equations have been using widely and the results have been reasonable so that "in spite of their limitations these methods will probably continue to be used" (Black et al., 1989).

Most of the researchers attempted to establish an equation which would be able to determine evaporation by using minimum number of meteorological parameters. Thornthwaite (1948) and Blaney and Criddle (1950) developed equations which included only air temperature. These equations are valid as well as useful for estimation of evapotranspiration in some areas so that FAO (Doorenbos and Pruitt, 1984) has recommended the Blaney-Criddle equation as one of the applicable method for estimation of consumptive use of water by plants. However, Doorenbos and Pruitt (1984) stated five limitation for using this equation, for example, he recommend that it should not be used in small islands and coastal areas where air is affected by sea temperature which has little response to seasonal change in radiation.

Most of the statistical models developed by applying measured data and consequently calibrated for the specific climates of that area. "The results confirm that an equation calibrated for one region is not, in general, valid for another geographical region and different climatic condition" (Jaywardena, 1989).

For this reason it would be desirable to test different equations in the study area and compare them to choose the best calibrated equation with the appropriate climatic conditions. Using the different equations and models to estimate evaporation in different climates needs the certain data which are not always available. In this case at first the necessary data must be produced and then applied in an appropriate evaporation equation. Nevertheless, in the most applications when more accurate estimation is essential and there is no limitation in the availability of meteorological data, the use of the theoretical method is preferable.

Shuttleworth (1983) classified different types of models used for estimation of evaporation and their data requirement. He investigated current usage of these models and concluded that the largest current usage belong to the Energy-Balance model.

Using Remote Sensing Data to Evaluate Evaporation

Using the pre-mentioned methods one can determine evaporation or evapotranspiration individually, in a field or local scale. However, evapotranspiration varies spatially and temporally because land surface and its characteristics changes especially and also temporally in large scale. Using Geographic Information Systems and Remote Sensing technology researchers have develop some spatial models to take into account spatial variability of evapotranspiration.

RS data can be used for measuring spatial distribution of land surface variables, such as vegetation cover, surface albedo and land surface temperature. Using these RS data in combination with topographic and atmospheric data, some physical and empirical models can be developed to estimate ET for clear sky conditions (Sanaeinejad & Hanna, 1994, Bastiaanssen, 2000; Su, 2002; Wang et al., 2006).

Different topographic and surface characteristics such as slope, aspect and vegetation cover, are needed for the spatial modeling of evaporation. Preparing the crude data for a GIS analysis involves a number of techniques. This procedure starts by digitizing a contour map of the area, processing satellite images and calculating some general hydrological features of the study area. This procedure on its own is mechanical, but time consumes, since original materials have to be prepared with a high degree of accuracy.

The models based on remote sensing data usually use the energy balance equation in which radiation components are the main parameters. Therefore the basic task in developing these models is determining radiation components over the study area.

SOLAR RADIATION MODELING

Insulation on a horizontal surface is given by the following equation:

$$R_H = R_0 \sin \gamma \tag{17}$$

in which R_0 is the solar constant and γ is the altitude of the sun and is equal to the angle between the solar beam and the insulated plane.

Using some spherical trigonometric equations the above equation can be converted in terms of the local geometric parameters (Robinson, 1966) as follows:

$$R_s = R_0[(\sin\varphi \sin\delta + \cos\varphi \cos\delta \cos t) \cos\varepsilon + \{ \cos A [\tan\varphi (\sin\varphi \sin\delta + \cos\varphi \cos\delta \cos t) - \sin\delta \sec\varphi] + \sin A \cos\delta \sin t\} \sin\varepsilon] \tag{18}$$

where R_s is the intensity of radiation on the surface inclined at various angles to the horizon, R_0 is the solar constant, ε is the slope of the surface, φ is the local latitude, δ is the declination of the sun, t is the time angle of the sun and A is the aspect of the surface measured from the south.

The integration of the above formula throughout a day from sunrise to sunset gives the daily average of direct radiation intensity on the surface.

$$R_{sd} = (24/\pi)\, R_0\, [T \sin\varphi \sin\delta \cos\varepsilon + \cos\varphi \cos\delta \sin T + (T \tan\varphi \sin\varphi \sin\delta + \sin T \cos\varphi \cos\delta \tan\varphi - T \sin\delta \sec\varphi) \sin\varepsilon \cos A - \cos\delta \cos T \sin A \sin\varepsilon] \quad (19)$$

where R_{sd} is the average daily solar radiation, T is the day length angle of the sun derived by:

$$\cos T = -\tan\varphi \tan\delta. \quad (20)$$

and the other symbols are the same as the equation (18). The declination of the sun for each day is calculated using the following equation (Paltridge and Platt, 1976):

$$\delta = 0.006918 - 0.399912 \cos\theta + 0.070257 \sin\theta - 0.006758 \cos 2\theta + 0.000907 \sin 2\theta - 0.002697 \cos 3\theta + 0.001480 \sin 3\theta. \quad (21)$$

The angle θ is defined in respect of the day number, d_n. Its value is zero on the first of January and count up to 364 on December 31. The equation is:

$$\theta = 2\pi d_n/365. \quad (22)$$

ESTIMATING GLOBAL RADIATION ON THE SURFACE

Equation (18) calculates the incident radiation on a non horizontal surface without considering the effect of the atmosphere. Since there are measured data of solar radiation available for the weather stations, a statistical method can be developed to take account the effect of atmosphere in this analysis.

There are two approaches to calculate solar radiation for different slopes and aspects.

1. First the equation (17) has been used to calculate the incident radiation in a horizontal surface. Then a simple regression model can be used to estimate monthly accumulative global radiation on a horizontal surface for each horizontal pixel.
2. The direct radiation for each individual cell can be calculated using equation (18), when the direct and diffuse components of global radiation in the watershed were estimated. The diffuse radiation also varies according to slope. This variation could also be considered in the model using the following equation (Wilson, 1970).

$$R_{ds} = \cos^2(\varepsilon/2)\, R_{dh}. \quad (23)$$

Where R_{ds} is diffuse radiation in a tilted surface, ε is the slope and R_{dh} is diffuse radiation on a horizontal surface. This allows the diffuse fraction of the global radiation to be estimated for the area and the remaining fraction is direct radiation which can be used to calculate direct radiation according to slope and aspect. And finally to calculate global radiation in each cell the two components of direct and diffused radiation are added together.

Modeling of Net Radiation over a Terrain Surface

The balance of incoming and outgoing radiation over a given period for a specific area can be defined as net radiation. Since there are different factors influencing net radiation, modeling this parameter is more problematic than global radiation. Several methods have been applied by different authors to calculate net radiation over a slope surface (Wilson and Garnier, 1975 and Varley, 1993). Frequently reflection of long wave radiation in a terrain surface makes the modeling more complex. However, if the measured data for various radiation components at a representative horizontal site is available, it is possible to evaluate the surface variation of radiation balance and then calculate the value for the surrounding terrain area.

Theory of Calculating Net Radiation

For a horizontal surface the energy balance equation can be written as follows:

$$R_n = (1-\alpha)R + L\downarrow - L\uparrow \qquad (24)$$

Where R_n is the net radiation, $\propto$ is the albedo, R is the global radiation, $L\downarrow$ is incoming long wave radiation from the atmosphere and $L\uparrow$ is the outgoing long wave radiation. Considering some other radiation components for a sloping surface a similar equation can be written for a terrain surface as follows:

$$R_{ns} = (1-\alpha)(R_s+R_r) + L_s\downarrow - L_s\uparrow + L_r + L_e \qquad (25)$$

In which the notations are the same as equation (24) with the subscript s referring to slopping surface. R_r is the reflected global radiation and L_r is the reflected long wave radiation onto a slope from the surrounding surfaces, L_e is the long wave radiation emitted by adjacent surfaces which is absorbed by the slopping surface.

A small amount of energy is involved in the four terms R_r, $L_s\downarrow$, L_r, and L_e. Wilson and Garnier (1975) indicated that this is not more than 3% for a 20° slope which is less than the energy received by a horizontal surface. They also found that the sum of the three terms (R_r, L_r, and L_e) would probably never exceed 14 Wm^{-1} for a vegetated surface during the summer. The term R_s can be calculated using equation (18) and α is assigned to the surface in terms of the vegetation cover. The term $L_s\uparrow$ cannot be calculated from the horizontal measurement data using physical laws and similar equations as the equation (18) (Wilson and Garnier, 1975). To determine this term, it should be either measured at each slopping surface or estimated using a regression equation. Therefore the practical equation to model net radiation over a terrain will be:

$$R_{ns} = (1-\propto)\, R_s + L_r \qquad (26)$$

Where L_r is the net long wave radiation.

Calculating Evapotranspiration Based on Energy Balance Method

Energy balance equation can be used for calculating evapotranspiration considering the surface characteristics. This is related to how dealing with the heat fluxes of vegetation and soil. In one case soil and vegetation covered surfaces are treated as a single energy source and in the other they are considered as two different radiation sources.

In these models only vertical energy transformation from the surface is considered and the required energy for photosynthesis is neglected. Evapotranspiration is controlled by the soil resistance to evaporate water and the stomata resistance in plants to transpiration. However, the problem is that the related data for these surface resistances are very hard to obtain. The lack of information can be compensated by satellite data. Remote sensing based models use the latent heat flux derived from the surface energy balance equation (Equation 1) (Moran *et al.*, 1994; Boegh *et al.*, 2002).

The instantaneous net radiation is also determined by surface characteristics, atmosphere properties, solar zenith angle, etc which was discussed earlier in this chapter. The surface emissivity and albedo are related to the vegetation cover which can be estimated by the Normal Difference Vegetation Index (NDVI) derived from the following equation:

$$NDVI = (NIR - R) / (NIR + R) \tag{27}$$

Where NIR is near infrared and R is red bands measured by satellite radiometric sensors (Chesapeaker Bay and Mid-Atlantic from Space, 2010). The sensible heat flux can be calculated by the following equation:

$$H = \rho C_p (aT_s + b)/r_a \tag{28}$$

In which a and b are experimental coefficients depends on the site characteristics, ρ is the air density (kgm^{-3}), C_p is the air specific heat at constant pressure ($Jkg^{-1}\ K^{-1}$), r_a is the aerodynamic resistance for heat transport($s\ m^{-1}$) calculated from the following equation (Katul et al, 1993).:

$$r_a = \frac{\ln\left(\frac{z-d_\circ}{z_{\circ m}}\right) \times \ln\left(\frac{z-d_\circ}{z_{\circ v}}\right)}{k^2 U_z} \tag{29}$$

Where z is the height at which wind and air temperature are measured; d_0 is the zero plane displacement height within the vegetation, z_{om} is the surface roughness length for momentum transport, z_{ov} is the surface roughness for vapor transport, k is von Karman constant of proportionality, 0.41, and U_z is the daily average wind speed.

d_0, z_{om} and z_{ov} are related to the vegetation height (h_c) by the following equations:

$$d_o = 0.67\ h_c \tag{30}$$

$$z_{om} = 0.123\ h_c \tag{31}$$

$$z_{ov} = 0.1\ Z_m \quad (32)$$

The crop resistance for heather and forest are taken to be $110sm^{-1}$ and $100sm^{-1}$ respectively (Hall and Harding, 1993, Varley 1993).

When sensible heat was calculated the evaporation can be obtained by the following equation:

$$E_f = \lambda E/(\lambda E+H) = (R_n - Gs - H)/(R_n - Gs) \quad (33)$$

Where E_f is the evaporation fraction in the energy balance equation. If we use a satellite image, then the energy balance equation gives the evaporation for one moment and needs to be extended over a day. The daily actual ET can be estimated from multiplying the evaporation fraction by the daily net radiation as follows (Shuttleworth et al., 1989; Crago and Brutsaert, 1996):

$$ET_a = (86400 \times E_f \times R_{day})/\lambda \quad (34)$$

Where ET_a is the daily actual evapotranspiration ($mmday^{-1}$), R_{day} is the daily net radiation ($W\ m^{-2}$).

There are a few algorithms that use the above models to estimate evapotranspiration based on RS data. Surface Energy Balance Algorithm for Land (SEBAL) (Bastiaanssen et al., 1998a), Surface Energy Balance System (SEBS) (Su, 2002), Parallel Two-Source Energy Balance (P-TSEB) (Shuttleworth and Wallace, 1985) and Series Two-Source Energy Balance (S-TSEB) (Kustas and Norman, 1999).

The GIS-RS evaporation models could provide reasonable spatial distribution of evapotranspiration across a wide range of land covers. This could be due to assimilating of remotely sensed surface temperature, which reflects the variation in evaporative (Batra et al., 2006). This technique can provide the key element that affect spatial variation of actual evapotranspiration. However, if we look at pixel scale there is some scepticism in the heat flux and evapotranspiration amount. This is because:

The accuracy of different parameters in land surface, which derived from remotely sensed data, is still under progress and has not achieved to a perfect level yet. The models based on remote sensing data usually use the energy balance equation. Each of the energy balance component can be derived reasonably by RS over the scene, but the absolute amount of latent heat flux may not absolutely adopt with reality.

Therefore, it is necessary to assess the accuracy of the RS based models carefully before being applied.

A Case Study for Spatially Analysis of Evaporation

Two watersheds were selected in Scotland for this study. A GIS model was developed to use RS data, weather station data, topographic data and Penman and Penman-Montieth equations for calculating evapotranspiration over the watersheds. In this model grass evaporation was estimated from Penman evaporation by applying the Hall and Harding

(1993) seasonal model. Transpiration for heather and forest was estimated using the Penman-Montieth equation applying aerodynamic resistance, r_a, from equation (29).

The most important spatially distributed variables in the radiation term were vegetation cover and net radiation. The effect of vegetation cover was derived from SPOT images and was applied in both of the radiation term and the crop resistance coefficient in the Penman-Montieth equation.

Net radiation pattern over the watershed was the main parameter to determine the radiation term of evaporation equations. The altitudinal effect on the aerodynamic term was also considered by applying the appropriate aerodynamic term for two different altitudinal zones.

Six different models were applied to calculate total evaporation for the watersheds and their results were compared. These are as follows.

1. The Penman equation using point measured data.
2. The GIS spatial model based on the Penman-Monteith equation.
3. The first version of Hall and Harding[1] (1993) model in which the Penman potential evaporation is used as an index to calculate potential plant transpiration for different vegetation cover and considering interception evaporation.
4. The second version of Hall Harding (1993) model in which the Penman-Monteith equation is used to calculate potential plant transpiration for different vegetation cover and considering interception evaporation.
5. The GIS spatial model based on the Penman equation.
6. The GIS spatial model based on the second version of Hall Harding (1993) model.

Technical procedure for a GIS spatial model: By using the weather station data, the coefficients for each of the terms in Penman and Penman-Montieth equation were calculated. Then the methods that was explained previously was applied in GIS environment to calculate the net radiation over the watersheds. The spatially calculated net radiation was used for calculating the radiation term of Penman and the Penman-Montieth equation for each cell over the watersheds.

The evaporation was calculated values for each cell and for each day individually, and daily evaporation values were accumulated to derive the monthly accumulation over the watersheds for the whole year. The results using Hall and Harding (1993) model were compared with the GIS spatial model just described.

The results of average Penman evaporation calculated using the GIS model for the watershed and the results based on the total average of evaporation derived from the single point measured data at the weather stations have shown in the Figure (1). The accumulation of average Penman-Montieth results is also shown for comparison. (The models 1, 2 and 5 as it was listed in the above)

The annual total accumulated evaporation is summarized in the Tble (1). It can be seen that the difference between the spatial GIS model of the Penman evaporation (model 5) and

[1] They used the Penman (1948) and Penman-Monteith equations (Montieth, 1975) to calculate evapotranspiration. Then using the following equation they calculated the overall catchment evaporation losses. $E = f_g E_g + f_h E_h + f_f E_f$ in which $f_g E_g$, $f_h E_h$ and $f_f E_f$, represent the evaporation fraction from grassland, heathland and forest area respectively. They considered the effect of altitude on evaporation by dividing the catchment into three altitudinal zones.

the values based on point measured data in the weather stations (model 1) is very small, approximately 2%, in Watershed 2. However, the difference is nearly 10% in Watershed 1 over one year.

The GIS model shows that topographic and vegetation features cause a variation in the spatial pattern of evaporation. Although the variation is small for daily evaporation, accumulation over a period of one year or more becomes significant. Thus, the difference between a spatially modeled Penman evaporation and a single site measurement strongly depends on the topographic condition and also on the variation of vegetation cover of the catchment.

Table 1. The annual accumulation of actual evaporation using the six models described in the text

	E-Act				Ep (point data)	Ep (Modelled)
	(Model 6)	(Model 4)	(Model 3)	(Model 2)	(Model 1)	Model (5)
Watershed 1	469.14	449.69	436.48	335.79	421.63	395.06
Watershed 2	533.87	527.14	381.27	434.11	485.82	475.69

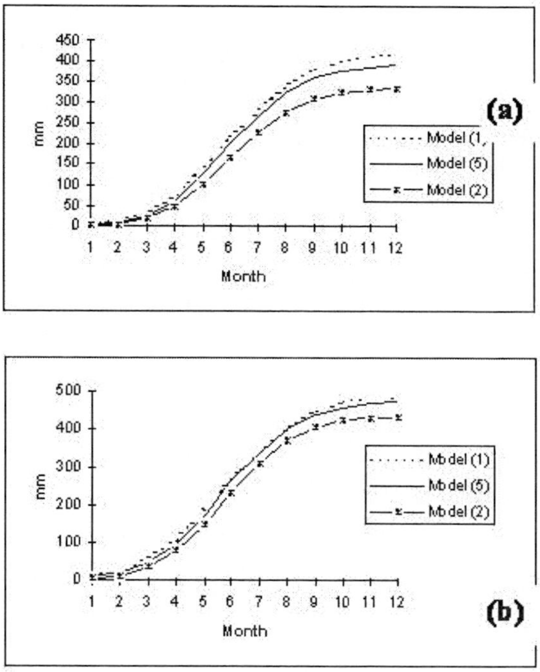

Figure (1). The monthly accumulation of Penman evaporation derived from the point data(model 1) compared with GIS modelled Penman (model 5) and the GIS modelled Penman-Montieth evaporation (model 2) in (a) watershed 1 and (b) watershed 2

Actual Evaporation

It is seen in the Figure 0-2a that total average of evaporation derived for Watershed 1 by the first version of Hall and Harding (1993) model (model 3) is very close to the values derived by the second version of Hall and Harding (1993) (model 4) and also with the GIS model developed in this study (model 6). This is because Hall and Harding (1993) considered transpiration from heather as $0.5E_p$ and the forest transpiration as $0.9E_p$ while the grass transpiration was considered as seasonally dependent. The percentage of heather is very small in Watershed 1 therefore the effect of the lower transpiration from heather does not affect the results for the spatial dependent model.

Figure 0-2b shows that there is a substantial difference between the second version of Hall and Harding (1993) model (Model 3) with the first version (model 4) and also with the GIS model (model 6). This difference can be explained in terms of the vegetation cover of the Watershed 2. Heather transpiration is underestimated in the first version of Hall and Harding (1993) (model 3) since low transpiration coefficient has been considered for heather in this model. More than 30% of the Watershed 2 is covered by heather therefore the low transpiration from this large heather land has a significant impact on total transpiration for the whole catchment.

To estimate the actual evaporation for the watersheds the transpiration in the second version of Hall and Harding (1993) model and also in the GIS model was calculated based on the Penman-Montieth equation. Therefore the difference between forest and heather transpiration was not as much as in the first version of the model.

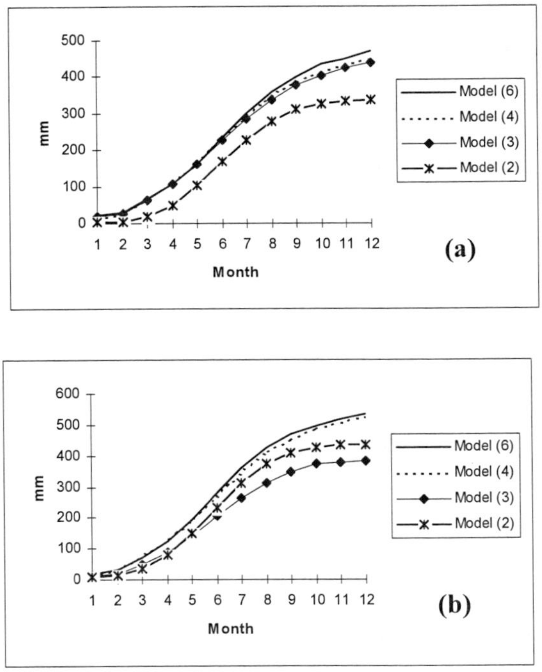

Figure 0-2. The monthly accumulation of actual evaporation using the GIS (model 6), the second version of H&H model (model 3) and the first version of H&H model (model 4) comparing with the Penman-Montieth evaporation modelled by GIS (model 2) in (a) Watershed 1 and (b) watershed 2

As it can be seen in the case of the Penman evaporation there is not a substantial difference between the actual evaporation estimated using the GIS model and the second version of the altitudinal zone model of Hall and Harding (1993) in Watershed. However, in Watershed 1 the actual evaporation estimated by GIS method was higher than Hall and Harding (1993) model (20mm in one year). This difference may not be significant in these particular watersheds, but the fact that the difference in Watershed 1 was higher than that in Watershed 2 showed that the effect of difference in topographic characteristics is distinguishable by the GIS model. This could make this model more generally applicable in a watershed with a variety of topography and vegetation.

Conclusion

Evaporation is a complicate process and has a major role in hydrology, meteorology, climatology, agriculture and other related disciplines. To identify the exact aspect of the evaporation study an exact definition of the related terminology is essential. Therefore the concepts of potential evaporation, potential evapotranspiration, actual evaporation and interception evaporation are exactly introduced.

Determination of evaporation has also been a problematic and difficult task. Different methods are used to determine evaporation depends on application and available data. Different methods for estimation of evaporation were discussed and some advantages and disadvantages were also explained. Finally, it was concluded that when there is enough meteorological data, the Penman equation is the best physical based method to estimate potential evaporation. The Penman - Monteith equation is also a developed version of this method considering the effect of plant physiological characteristic in transpiration process. Using these equations in combination with GIS and RS, it is possible to provide a spatially distributed assessment of evaporation over an area with different topographic shapes and also different vegetation cover.

References

Balek, J. (1977). Hydrology and Water Resources in Tropical Africa, *Elsevier*, Czechoslovakia.

Barry, R. G. (1973). Evaporation and transpiration. in McBoyle, G. *Climate in review*, ****

Bastiaanssen WGM. (2000). SEBAL-based sensible and latent heat fluxes in the irrigated Gediz Basin, Turkey. *Journal of Hydrology*, *229*, 87-1000.

Batra, N., Islam, S., Venturini, V., Bisht, G. & Jiang, L. (2006). Estimation and comparison of evapotranspiration from MODIS and AVHRR sensors for clear sky days over the Southern Great Plains. *Remote Sensing of Environment*, *103*, 1-15.

Black, J. N., Bonython, C. W. & Prescott, J. A. (1954). Solar radiation and the duration of sunshine. *Quart. J. R. Met. Soc.*, *80*, 231.

Black, T. A., Splittlehouse, D. L., Novak, M. D. & Price, D. T. (1989). *Estimation of areal Evapotranspiration*. LAHS Publication No.177, Galliard Ltd, Great Yarmouth.

Blaney, H. F. & Criddle, W. D. (1950). *Determining water requirements in irrigated areas from climatological and irrigation data.* USDA soil Conserv. Soil Conserv. Serv. SCS TP 96 (processed) *44*.

Brisson, N. & Seguin, B. (1992). Meteorological Soil Water Balance for Crop Simulation Models . *Agric. For. Meteor.*, *59*, 267-287.

Brisson, N. & Seguin, B. (1992). Meteorological Soil Water Balance for Crop Simulation Models . *Agric. For. Meteor.*, *59*, 267-287.

Bruijnzeel, L. A. & Wiersum, K. F. (1987). Rainfal interception by a young Acacia auriculiformis(A.Cunn) plantation forest in west Java, Indonesia: Application of Gash's analytical model. *Hydrol. Process.*, *1*, 309-319.

Brunt, D. (1939). *Physical and dynamic meteorology*. Cambridge University. Press.

Brutsaert, W. h. & Stricker, H. (1979). An advective-aridity approach to estimate actual evapotranspiration. *Water Resour. Res.*, *15*, 443-450.

Brutsaert, W. (1982). Evaporation into the atmosphere, theory, *history and application.*, Reidel Pub. Co., Boston.

Calder, I. R. (1986). A stochastic model of rainfall interception. *J. Hydrol.*, *89*, 65-71.

Calder, I. R. (1976). The Measurement of Water Losses from a Forested Area Using a "Naturnal" Lysimeter . *J. Hydrol.*, *30*, 311-325.

Calder. I. R. (1990). Evaporation in the uplands. *Chichester*, New York.

Chesapeaker Bay and Mid-Atlantic from Space, (2010). http:// chesapeake. towson. edu/ data/ all_ndvi.asp

Crago, R. & Brutsaert, W. (1996). Daytime evaporation and the selfpreservation of the evaporative fraction and the Bowen ratio. *Journal of Hydrology*, *178(1-4)*, 241-255.Bastiaanssen et al., 1998a.

Dalton, J. (1802). On evaporation. Essay III in: Experimental essays on the constitution of mixed gases; on the force of steam or vapour from water or other liquids in different temperatures; both in a Torrecellian vacuum and in air; on evaporation; and on the expansion of gases by heat. *Mem. Proc. Lit. Phil. Soc.*, Manchester, 5, part *2*, 575-595.

Denmead, O. T. & Show, R. H. (1962). Availability of soil water to plants as affected by soil moisture content and meteorological conditions. *Agron. J*, *45*, 385-390.

Dickinson, R. E., Henderson-Sellers, A., Rosenweig, C. & Sellers, P. J. (1991). Evapotranspiration models with canopy resistance for use in climate models, a review. *Agric. For. Meteor.*, *54*, 373-388.

Dickinson, R. E., Henderson-Sellers, A., Rosenweig, C. & Sellers, P. J. (1991). Evapotranspiration models with canopy resistance for use in climate models, a review. *Agric. For. Meteor.*, *54*, 373-388.

Doorenbos, J. & Pruitt, W. O. (1984). Crop Water Reqirement. FAO No: 24, Rome Italy.

Gash, J. H. C. & Morton, A. J. (1978). An application of the Rutter model to the estimation of the interception loss from Thetford Forest. *J. Hydrol*, *38*, 49-58.

Gash, J. H. C. & Morton, A. J. (1978). An application of the Rutter model to the estimation of the interception loss from Thetford Forest. *J. Hydrol*, *38*, 49-58.

Gash, J. H. C. & Stewart, J. B. (1977). The Evaporation from Thetford Forest During 1975. *J. Hydrol.*, *35*, 385-396.

Giambelluca, T. W. & Nullet, D. (1992). Evaporation at High Elevations in Hawaii. *J. Hydrol.*, *136*, 219-235.

Granger, R. J. (1989). An examination of the concept of potential evaporation. *J. Hydrol.*, *111*, 9-19.

Granger, R. J. (1989). An examination of the concept of potential evaporation. *J. Hydrol.*, *111*, 9-19.

Green, F. H. W. (1970). Some Isopleth Maps Based on Lysimeter Observation in the British Isles in 1965, 1966, 1967 . *J. Hydrol.*, *10*, 127-140.

Hall, R. L. (1987). Processes of Evaporation from Vegetation of the Uplands of Scotland. Transactions of Royal Society of Edinburgh: *Earth Science*, *78*, 327-334.

Hanna. L. & Siam, N. (1981). The empirical relation between sunshine and radiation and its use in estimating evaporation in North East England. *J. Clim.*, *1*, 11-19.

Hellwig, J. (1973). Evaporation of water from sand, diurnal variation. *J. Hydrol*, *18*, 109-118.

Helvey, J. D. (1971). A summary of rainfall inteception by certain conifers of Noth America. I: Biological effects in the hydrological cycle. In: Third Int. *Seminar for Hydrology Professors Proc.*, Prude University, West Lafayett, Indiana. 103-113.

Jaywardena, W. A. (1989). Calibration of some empirical equations for evaporation and evapotranspiration in Hong Kong. Agric. *For. Meteor*, *47*, 75-81.

Jensen, M. E., Burman, R. D. & Allen, R. G. (1990). *Evapotranspiration and irrigation water requirements.* American Society of Civil Engineers, New York.

Johnson, R. C. (1990). The Interception, Throgh fall and Stem flow in a Forest in Highland Scotland and the Comparison with Other Upland Forests in the UK. *J. Hydrol.*, *118*, 281-287.

Jones, K. R. (1981). *Arid Zone Hydrology.* FAO Irrigation And Drainage Paper No. *37*, Rome, Italy.

Kustas, W. P. & Norman, J. M. (1999). Evaluation of soil and vegetation heat flux predictions using a simple two-source model with radiometric temperature for partial canopy. *Agricultural and Forest Meteorology*, *94*, 13-29.

Lafleur, P. M. (1992). Energy balance and evapotranspiration from a subarctic forest. *Agric. For. Meteor.*, *58*, 163-175.

Lemeur, R. & Zhang, Iu. (1990). Evaluation of Three Evapotranspiration Models in Terms of Their Applicability for an Arid Region. *J. Hydrol.*, *114*, 395-411.

Lloyd, C. R., Gash, J. H. C., Shuttleworth, W. J. & Marques, F. A. de O. (1988). The measurement and modelling of rainfall interception by Amazonian rain forest. *Agric. For. Meteor.*, *43*, 277-294.

Makkink, G. F. (1957). Testing the Penman formula by means of lysimeters. *J. Inst. Water Engng.*, *11(3)*, 277-288.

Malek, E. (1992). Night-time evapotranspiration vs. daytime and 24h evapotranspiration. *J. Hydrol.*, *138*, 119-129.

Mcllory, I. C. & Angus, D. E. (1964). Grass, water and soil evaporation at Aspendale. *Agric. Meteor*, *1*, 201-224.

Mondeville, A. N. & Batchelor, C. H. (1990). Estimation of actual evapotranspiration in Malawi. *Inst. Hydrol. Rep.*, No. *110*.

Monteith, J. L. (1965). *Evaporation and the environment. In the state and movement of water in living organism, XIXth Symposium.* Soc. for Exp. biol., Swansea, Cambridge University Press.

Moran, M. S., Kustas, W. P., Vidal, A., Stannard, D. I., Blanford, J. H. & Nichols, W. D. (1994). Use of ground-based remotely sensed data for surface energy balance evaluation of a semiarid rangeland. *Water Resources Research, 30(5)*, 1339-1349. Boegh et al., 2002

MTSAT-1R, (2010). http://www.jma.go.jp/jma/jma-eng/ satellite/ Basic_ Information/ Basic_ Information.html

Mulder, J. P. M. (1983). A simulation of rainfall interception in a pine forest. Dissertation thesis, University of Groningen, Netherland.

Osmar, A. C. & Cuenca, R. C. (1992). Precision of Evapotranspiration Estimates Using Neutron Probe. *J. Irrig. Drain. Engin., 118*, 6, 943-953.

Paralange, M. B. & Katul, G. G. (1992). Estimation of the diurnal variation of potential evaporation from a wet bare soil surface. *J. Hydrol., 132*, 71-89.

Parlange, M. B. & Katul, G. G. (1992). Estimation of the diurnal variation of potential evaporation from a wet bare soil surface. *J. Hydrol., 132*, 71-89.

Pasquill, F. (1943). Evaporation from a plane, free liquid surface into a turbulent air stream., *Proc. Roy. Soc.*, London, A. *182*, 75-95.

Penman, H. L. (1948). Natural evaporation from open water, bare soil and grass. *Proc. R. Soc.*, London, *Ser. A. 193*, 120-146.

Persson, G. & Lindroth, A. (1994). Simulating evaporation from short-rotation forest. *J. Hydrol, 156*, 21-45.

Prescott, J. A. (1940). Evaporation from a water surface in relation to solar radiation. *Trans. R. Soc. S. A, 64*.

Priestley, C. H. B. & Taylor, R. J. (1972). On the assessment of surface heat flux and evaporation using large scale parameters. *Mon Wea. rev., 100*, 81-92.

Ritchie, J. T. (1972). Model for predicting evaporation from a row crop with incomplete cover. *Water Resour. Res., 8(5)*, 1204,1213.

Robinson, N. (1966). *Solar radiation*. Elsevier Co. Amsterdam.

Rowher, C. (1931). Evaporation from free water surfaces. *USDA Tech. Bull., 271*. 96.

Rowntree, P. R. R. (1988). Atmospheric parameterization schemes for evaporation over land: Basic concepts and climate modeling aspects. In Schmugge, T. J. & Andre, J.C. *Land surface evaporation*, Spring-Verlage, New York.

Running, S. W. (1991). Computer simulation of regional evapotranspiration by integrating landscape biophysical attributes with satellite data. T. J. Schmugge, & J. C. Andre, (Editors), *Land surface evaporation, measurement and parameterization*. Springer-Verlag, New York.

Rutter, A. J. (1963). Studies in the water relations of Pinus Sylvestrris in plantation conditions. In Measurements of rainfall and interception, *J. Ecol., 51*, 191-203.

Rutter, A. J. (1967). An analysis of evaporation from a stand of Scots pine. In: W. E. Sopper, & H. W. Lull, (Editors), *Int. Symp. on Forest Hydrology*, Pergamon Press, Oxford, 403-417.

Sanaeinejad & Hanna, 1994.

Sanaei-Nejad, S. H. & Hanna, L. W. (1995). Simulating of the solar radiation over the Balquhidder watershed in centeral Scotland. *The summer meeting of the climatological group of Royal Meteorology Association*, Sheffild, UK.

Schmugge, T. & Andre, J. C. (1991). HYPEX-MOBILHY: Results from a large-scale experiment. Remote Sensing and large-scale Global Processes. IAHS Publication No. *186*, *Walingford*, England.

Shih, S. F., Allen, L. H., Hammond, L. C., Jones, J. W., Rogers, J. S. & Smajstral, A. G. (1983). Basinwide water requirement estimation in southern Florida. Trans. *ASAE, 26 (3)*, 760-766.

Shih, S. F. & Cheng, K. S. (1991). Evapotranspiration estimation for data-short environment. *J. Irrig. Drain. Engin., 117*, 1, 107-122

Shuttleworth, W. J., Gurney, R. J., Hsu, A. Y. & Ormsby, J. P. (1989). Fife: *The variation in energy partition at surface flux sites*. IAHS Publication, *186*, 67-74.

Shuttleworth, W. J. & Wallace, J. S. (1985). Evaporation from sparse cropsan energy combination theory. *Quarterly Journal of the Royal Meteorological Society, 111*, 839-855.

Shuttleworth, W. J. (1983). Evaporation models in the global water budget. in Perrott, A. M., Beran, M. & Ratcliffe, R. *Variation in the global water budget. Reidel*, D. Publishing Co. Lancaster.

Siam, N. (1979). *SoilMoisture changes due to evapotranspiration and drainage under arable land. PhD thesisi*, University of Newcastle Upon Tyne.

Smith, W. R. (1967). Potential transpiration. Technical bulletin. No. 16, Min. of Agr. *Fish. and Food. (MAFF)*, London. Her Maj. Stationery Office.

SRCE, (2010). http://lasp.colorado.edu/sorce/

Stewart, J. B. (1977). Evaporation from the wet canopy of a pine forest. *Water Res., 13*, 915 - 921.

Stringer, E. T. (1972). *Techniques of Climatology*. W.H. Freeman and Company, San Francisco.

Su, Z. (2002). The surface energy balance system (SEBS) for estimation of turbulent fluxes. *Hydrology and Earth System Sciences, 6(1)*, 85-99.

Suckling, P. W. (1980). The Energy Balance Micro climate of Suburban Lawn. *J. Appl. Meteor., 19*, 607.

Sutton, (1934). Wind structure and evaporation in a turbulent atmosphere. *Proc. Roy. Soc. London A., 148*, 701-722.

Thornthwaite, C. W. (1948). An approach toward a rational classification of climate. *Geograph. Rev., 38*, 55.

Van Bavel, C. H. M. (1953). A drought criterion and its application in evaluating drought incidence and hazard. *Agron. J. 45*, 167-172.

Van Bavel, C. H. M. (1966). Potential evaporation: the combination concept and its experimental verification. *Water Resour. Res., 2*, 455-467.

Varley, M. (1993). *A method of predicting spatial distribution of evaporation in ypland areas from single point meteorological measurement*. Ph.D thesis. Lancaster University, UK.

Veihmeyer, F. J. (1927). Some factors affecting the irrigation requirements of deciduous orchards. *Hilgardia, Vol. 2*, 125-284.

Wang, K., Li, Z. & Cribb, M. (2006). Estimation of evaporative fraction from a combination of day and night land surface temperatures and NDVI: A new method to determine the Priestley-Taylor parameter. *Remote Sensing of Environment, 102*, 293-305.

Wilson, J. I. B. (1970). *Solar energy*. Taylor and Francis Ltd.

Wilson, R. G. & Garnier, B. J. (1975). Calculateed and measured net radiation for slope. *Climate Bull., Vol. 17*, 1-14.

Wright, I. R. & Harding, R. J. (1993). Evaporation from Natural Mountain Grassland. *J. Hydrol.*, *145*, 267-283.

Zhanqing, Li. & Leighton, H. G. (1993). Estimation of SW flux Absorbed at the Surface from TOA Reflected Flux. *J. Climat.*, *317*.

In: Relative Humidity: Sensors, Management…
Editor: John M. Castillo

ISBN: 978-1-61761-734-8

Chapter 3

HUMIDITY-SENSITIVE AND NON-OHMIC TIN DIOXIDE CERAMICS

A.B. Glot*

División de Estudios de Posgrado, Universidad Tecnológica de la Mixteca,
Huajuapan de León, Oaxaca, México

ABSTRACT

Oxide ceramic semiconductors obtained by sintering of tin dioxide (SnO_2) with small additives of other oxides, exhibit strong deviations from Ohm´s law (varistor effect) and strong humidity dependence of conductivity at low electric fields (humidity sensor effect). SnO_2-based ceramics are suitable for the preparation of a device with double function (varistor and humidity sensor). Electrical conduction in such n-type semiconductors is controlled by the grain-boundary potential barriers formed as a result of sintering in oxidizing atmosphere. Humidity-sensitive and non-Ohmic properties of SnO_2 ceramics are explained by the decrease of the barrier height with the relative humidity and the electric field, respectively. Starting from these assumptions the functional dependences of electrical conductivity on relative humidity and on electric field are calculated and compared with experimental data. It is shown that electrical conductivity exponentially strongly depends on the relative humidity and the electric field.

Keywords: electrical conduction, grain boundary, humidity sensor, non-Ohmic conduction, potential barrier, relative humidity, tin dioxide, varistor.

1. INTRODUCTION

Electrical properties in various ceramic oxide materials exhibit dependence on the relative humidity of air and these effects are used for the development of humidity sensors [1-

* Corresponding author: e-mail: alexglot@mixteco.utm.mx, fax: + 52 (953) 5320214 ext 106.

3]. Various aspects of the ceramic sensors for humidity detection are described in the literature and summarized in many reviews (see, for example [4-8]). Most frequently the influence of humidity on electrical resistance and capacitance are reported. Respectively, resistive-type humidity sensor is a device with the dependence of its resistance on relative humidity. Resistive-type humidity sensors can be classified into two types: ionic and electronic [2]. In the first type, porous ceramic elements made with insulating materials exhibit protonic (H+) conductivity in a humid atmosphere due to capillary condensation of water vapour, however, the humidity-sensitive characteristics depend upon the conductivity and the microstructure of the material (in particular, pore size distribution) [2,9]. The second type utilizes the sensitivity of some semiconducting materials to water vapour. The chemisorption of water molecules increases or decreases the electronic conductivity depending on the n- or p-type of conduction [2].

The interaction of water vapour with oxide surface causes complex physico-chemical processes at the surface strongly dependent on the nature of oxide adsorbent [2,10]. Therefore, several mechanisms for the explanation of the humidity sensor effect in different oxide ceramics were discussed [2-10] and search of new humidity-sensitive materials is continued [8,11-14]. Different inhomogeneous structures with nonlinear current-voltage characteristics can be of some interest not only for the gas detection [15] but for the humidity detection as well [7,14,16-18].

Tin dioxide based SnO_2-ZnO-Bi_2O_3 ceramics [19] and SnO_2-ZnO ceramics with other additives [20] exhibit nonlinear current-voltage characteristics (the effect of non-Ohmic conduction or varistor effect) and sensitivity of their electrical properties to the atmosphere of heat treatments. It was observed that electrical conductivity in tin dioxide based SnO_2-ZnO-CoO-Bi_2O_3 (SiO_2, GeO_2) ceramics depends not only on the electric field but on the relative humidity of air (humidity sensor effect) [21]. Both these phenomena are due to the fact that electrical conduction in tin dioxide based ceramics is controlled by the grain-boundary potential barriers formed during sintering in oxidizing atmosphere [21]. Humidity and electric field dependence of conductivity is rather common effect in SnO_2-based ceramics because it was shown recently that in various SnO_2-based ceramics with strong non-Ohmic behaviour at high fields, low-field conductivity is highly sensitive to the variation in the relative humidity of air [22-25]. SnO_2-based semiconductor ceramics are materials with double functions: varistor and humidity sensor [21-25]. Humidity dependence of conductivity takes place as well in ZnO varistors [17] and in sandwich-structures which are the models of grain boundaries in ZnO varistors [16,17], therefore, humidity and field dependence of conductance is not limited only by tin dioxide ceramics.

The dependences of electrical conductivity in oxide ceramics on electric field and on relative humidity are usually considered independently due to the difference in the nature of an external influence (electric field and surrounding atmosphere). Meanwhile, both effects can be analysed by similar way due to they have the same reason – decrease in the barrier height. It would be interesting to consider both mentioned effects in the same materials using similar approaches based on a decrease in the barrier height observed in electric field and in humid air. Recently similar approach to the understanding of non-Ohmic conduction in oxide varistor ceramics was developed [26-28] but humidity dependence of the electrical conductance was not discussed.

Therefore, the purpose of this chapter is to obtain the equations for the dependences of the electrical conductivity on the relative humidity and on the electric field in a system where the grain-boundary potential barriers control electrical conduction (for example, in SnO_2 semiconductor ceramics) using general approach based on the decrease of the barrier height in humid atmosphere and in electric field.

2. Experimental Details

Tin dioxide varistor (or varistor-sensor) ceramics are functional materials obtained by ceramic technology: mixing of tin dioxide (SnO_2) as a main ingredient with some additives, compacting of mixed powder and sintering at high temperature in oxidizing atmosphere. SnO_2 varistor-sensor ceramics are n-type semiconductors. The detailed description of the preparation procedure can be found in [21,22,24] but some typical procedure was the next. Oxide additives were in the range 0.05-1 mol.%. The weighted oxides were mixed in distilled water and dried. Obtained powder was pressed under axial pressure about 40-50 MPa in tablets about 10 mm in diameter and about 1 mm thick. Then pressed tablets were sintered in air by slow heating (at about 400 K/hour) up to temperature about 1470 K. This temperature was constant during 1 hour and then the furnace was slowly cooled to room temperature with the same rate. After grinding of a sintered sample without using water a suitable silver paste was overlapped at flat surfaces and a sample was slowly heated in air up to temperature 1070 K and after 10 minutes at this temperature it was slowly cooled.

Current-voltage characteristics were recorded by applying d.c. voltage and measuring the steady-state current. The results were obtained avoiding the self heating of the samples. The capacitance and ac conductance were measured at frequency 1 kHz. Electrical measurements were performed in air with different fixed relative humidity in the range 10 – 93 %. A sample was placed in a closed chamber above the surface of water solution of a proper salt for a time of about 1 hour before starting a measurement.

The nonlinearity of current-voltage characteristic can be characterized by the empirical nonlinearity coefficient defined as:

$$\beta = (U / I)(dI / dU) = d(\ln I) / d(\ln U), \tag{1}$$

where U is the voltage and I is the current. The nonlinearity coefficient can be defined as well using the current density j and the electric field E ($\beta = (E / j)(dj / dE)$). It gives the slope of current-voltage characteristic presented in double logarithmic scale. The nonlinearity coefficient depends on voltage (or current). Sometimes it is necessary to have only one numerical parameter for the characterization of varistor. In such cases the value of the nonlinearity coefficient, for example, calculated at the current density 10^{-3} A cm^{-2} is used.

3. Humidity and Electric Field Dependence of Conductivity in Oxide Ceramics

In this section the experimental data on the dependence of electrical conductivity on relative humidity and electric field for several ceramic materials are presented.

It is necessary to make the important remark. It was assumed [21] and later shown experimentally [23] that in humid air peripheral region of a sample becomes more conductive than the central one. A sample becomes electrically inhomogeneous. It is correct in respect to a sample placed in humid air to use the values "the current" and "the conductance" ($G = I / U$). Such specific physical values like "the current density" ($j = I / S$, where S is the cross-section of a sample) and "the conductivity" ($\sigma = I \cdot d / U \cdot S$, where d is the sample thickness) looks estimated not very precisely. However, it is convenient to use specific values in the calculations with the intention to avoid unnecessary complications in formulas. Therefore, with such reservation below in the calculations the specific physical values (j and σ) will be used but in the discussion of experimental data "the conductance" will be used as well. This should not cause any difficulty as far as it is always possible to convert one into another.

3.1. $MgCr_2O_4$-TiO_2 Ceramics

The ceramics in the system $MgCr_2O_4$-TiO_2 are used as humidity sensors [1,4]. Ceramic materials are characterized by their structure consisting of crystal grains, grain boundaries, grain surfaces and pores. The controlled-pore ceramics are suitable for applications as atmospheric sensors such as humidity and gas sensors [1,4]. Since metal oxides with high Cr content are generally difficult to sinter to high density, the sintered product tends to exhibit a typical porous structure [4]. The $MgCr_2O_4$-TiO_2 ceramics display high mechanical and thermal properties at high temperature. The content of oxides in this system can be different. For example, in [1] the material with composition (mol.%) 39.4 MgO, 39.4 Cr_2O_3, 21.2 TiO_2 was studied, though, in [4] the content of TiO_2 was 35 mol.%.

The $MgCr_2O_4$-TiO_2 material (35 mol.% TiO_2) sintered at 1630 K (2 h) is single-phase with pure $MgCr_2O_4$–type spinel structure and developed porous structure with porosity about 25% and pour size 0.05-0.3 μm at the average grain size 1 μm [4]. The intergranular pores can be considered as a network of capillary tubes [4]. The $MgCr_2O_4$-TiO_2 ceramics are p-type semiconductors.

The adsorption below 420 K on porous $MgCr_2O_4$-TiO_2 material involves both physisorbed and chemisorbed water, though, at higher temperatures between 570 and 770 K reversible chemisorption becomes dominant [4]. Initially, a few water molecules chemisorb with formation of hydroxyl (OH) group. The hydroxyl groups adsorbing on the most active surface Cr^{3+}ions dissociate providing mobile protons (H^+) [4]. More water vapour adsorbs physically on the hydroxide surfaces to form hydroxyl multilayer. Therefore, the mechanism of surface conduction is ionic and protons are probably the dominant charge carriers [4]. Electrical conductance G in $MgCr_2O_4$-TiO_2 ceramics strongly depends on the relative humidity w (Figure 1) [4], though, quite noticeable deviations from linear relation between $\log G$ and w is seen. At high humidity some "saturation" takes place. This "nonlinearity" can be due to the specific pore-related conduction mechanism in humid air [2,4,7] which is different from the conduction mechanism related to the grain-boundary potential barriers (see [21-25]).

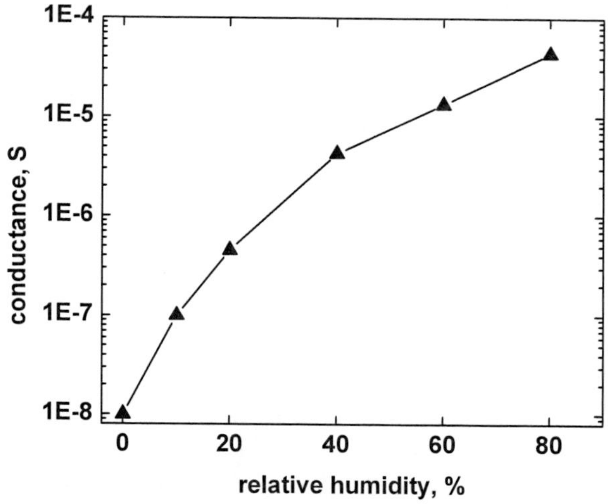

Figure 1. Dependence of conductance in $MgCr_2O_4$-TiO_2 ceramics on relative humidity. Replotted using data from [4]

$MgCr_2O_4$-TiO_2 ceramics probably do not exhibit substantial nonlinearity of current-voltage characteristic (at least such data were not reported in [4]).

3.2. SnO_2–Based Ceramics with Non-Ohmic and Humidity-Dependent Conduction

It would be interesting to compare humidity dependence of conductivity in $MgCr_2O_4$-TiO_2 ceramics with those observed in materials where grain boundaries control the conduction. In Figure 2 current-voltage characteristics for the SnO_2-ZnO-CoO-SiO_2 ceramic sample at two values of the relative humidity plotted using the data from [21] are presented. Increase in relative humidity causes displacement of nonlinear current-voltage characteristic to higher currents. Stronger shift is observed for low voltages (Figure 2).

As it is seen from Figure 2 the slope of current-voltage characteristic in double logarithmic scale $\log I - \log U$ is not a constant value: the nonlinearity coefficient is increased with voltage. It is clear from Figure 2 that at higher voltage the humidity sensitivity (the difference in current values calculated at fixed voltage) becomes lower and at high relative humidity the nonlinearity coefficient (the slope of $\log I - \log U$ dependence) is decreased. For the explanation of these facts one can assume that in humid air the peripheral area of the sample becomes more conductive and works as a linear electrical shunt to the bulk (resistance of this shunt is strongly dependent on relative humidity and at fixed relative humidity the increase in voltage can lead to the dominating of non-Ohmic bulk current over linear surface current) [21]. In general this explanation is correct. The strong redistribution of current in favour of the surface area was proved by direct experiment in [23].

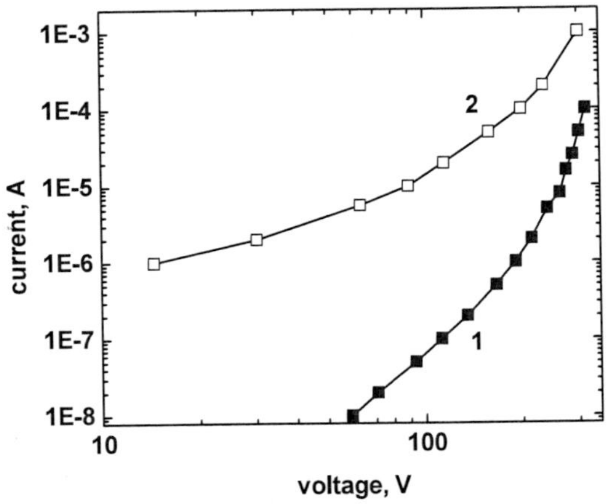

Figure 2. Dependences of current on voltage for SnO_2-ZnO-CoO-SiO_2 ceramics obtained at relative humidity 10% (1) and 90 (2). Replotted using data from [21]

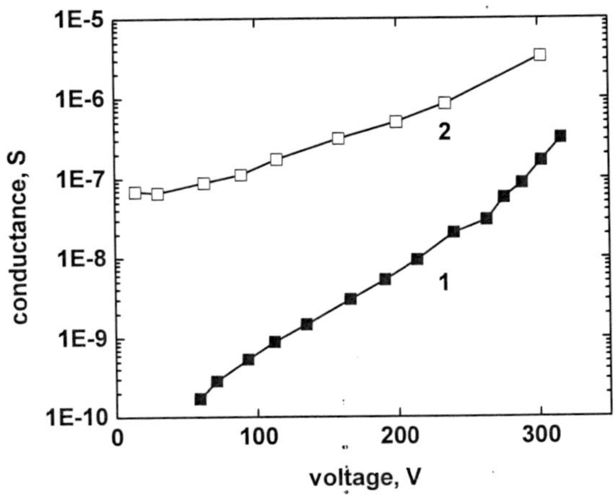

Figure 3. Dependences of conductance on voltage for SnO_2-ZnO-CoO-SiO_2 ceramics at relative humidity 10% (1) and 90% (2). Replotted using data from [21]

It was shown earlier that current-voltage characteristics of oxide varistor ceramics are linearized in the scale $\log\sigma - E$ (where σ is the electrical conductivity at electric field E, $\sigma = \sigma(E) = j(E)/E$) because the dependence of the current density on the electric field is given by the expression [26-28]:

$$j = \sigma_0 E \exp(\alpha E) \tag{2}$$

where σ_0 is the electrical conductivity at low electric field (Ohmic electrical conductivity). In the case of highly nonlinear ceramics (ZnO or SnO_2 varistors) two straight lines in the scale $\log\sigma - E$ are appeared [26-28]. Therefore, data from Figure 2 were plotted in the scale $\log G - U$ (where G is the conductance) and are presented in Figure 3. Here the voltage dependences of conductance at two relative humidity values are shown. They are linearized in the scale $\log G - U$ and the slope at higher humidity is decreased (Figure 3).

The dependence of conductance at fixed voltage on relative humidity is shown in Figure 4 (curve 1). It is nearly linear in the scale $\log G - w$ (Figure 4, curve 1) in comparison with clear tendency to saturation in $MgCr_2O_4$-TiO_2 ceramics (Figure 1).

The results of scanning electron microscopy of studied SnO_2-based ceramics are shown in Figure 5. In Figure 5 (a) the fracture surface of one of the first varistor ceramics [19] (with the nonlinearity coefficient $\beta \cong 10$) is presented. It is seen that in spite of Bi_2O_3 addition this material is quite porous. Two other images are related to SnO_2-based ceramics with the nonlinearity coefficient $\beta \cong 30-40$ (b) and $\beta \cong 40-50$ (c). These secondary electron images show quite porous structure of SnO_2 varistor-sensor ceramics. Some increase in density can be reached by Bi_2O_3 addition. (Figure 5 (c)) though bismuth oxide rather insufficiently wets surface of SnO_2. Higher densification was achieved by CuO addition [25,29].

In tin dioxide semiconductor ceramics the electrical conduction is controlled by the potential barriers formed at the grain boundaries during sintering in oxidizing atmosphere [21-23]. The increase in conductance of the surface area and the whole sample in humid air is attributed to the lowering of the barrier height because not only the conductance but the capacitance as well is increased with the rise of humidity (Figure 6) [30]. The increase in capacitance in Figure 6 can be due to the narrowing of the grain-boundary depleted region due to the decrease in the barrier height.

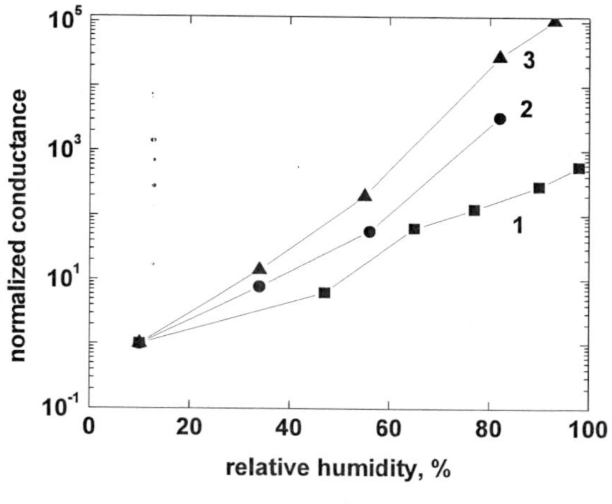

Figure 4. Dependences of normalized conductance on relative humidity for different SnO_2-based ceramics: 1 - SnO_2-ZnO-CoO-SiO_2 [21], 2 - SnO_2-Co_3O_4-Nb_2O_5-Cr_2O_3-Bi_2O_3 [23], 3 - SnO_2-Co_3O_4-Nb_2O_5-Cr_2O_3 [24]. Values of conductance at each relative humidity were normalized on the value of conductance at relative humidity 10%. Replotted using data from [21,23,24]

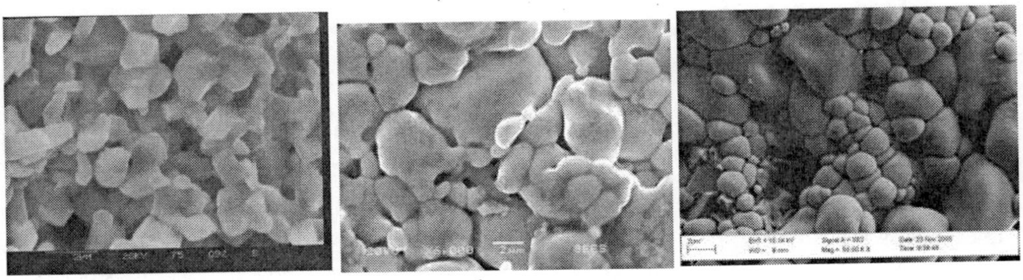

a b c

Figure 5. SEM micrographs of varistor-sensor ceramics SnO_2-ZnO-Bi_2O_3 (a), SnO_2- Co_3O_4-Nb_2O_5-Cr_2O_3 (b), SnO_2- Co_3O_4-Nb_2O_5-Cr_2O_3-Bi_2O_3 (c). The magnification of each image is nearly the same (the horizontal size of each picture is about 20 μm).

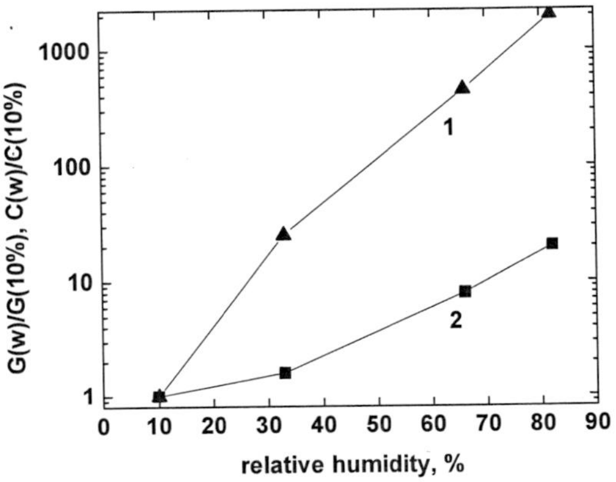

Figure 6. Relative ac (1 kHz) conductance (1) and relative capacitance (2) vs relative humidity for SnO_2-ZnO-GeO_2-Co_3O_4-Sb_2O_5 ceramics. Replotted using data from [30].

Last time a great deal of attention was focused on highly nonlinear current-voltage characteristics observed in SnO_2–based ceramics with some additives [31-35] which is comparable with the nonlinearity in ZnO varistors. In this case increase in electric field causes relatively weak current density rise followed by the strong increase in current density at high electric fields. The example of such behaviour is shown in Figure 7 (curve 1). Actually, curve 1 in Figure 7 was obtained in air with low relative humidity ($w = 10$ %). It was observed that increase in relative humidity leads to the shift of low-nonlinear part of current-voltage characteristic to higher current density but highly-nonlinear part of current-voltage characteristic is practically unchanged (Figure 7). In other words, highly-nonlinear current-voltage characteristic is observed even in quite humid air. This is because the increase in current at high electric field due to non-Ohmic conduction becomes significantly higher than the increase in current due to rise in relative humidity. This situation is similar to what was found for ceramics with lower nonlinearity (Figure 2) but expressed more pronouncedly.

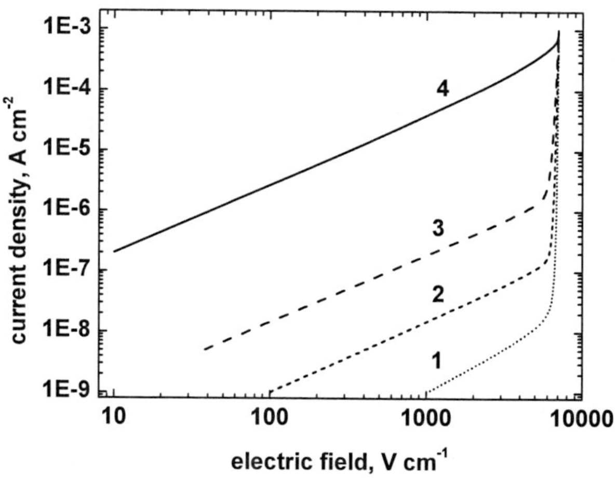

Figure 7. Current-voltage characteristics of SnO_2-Co_3O_4-Nb_2O_5-Cr_2O_3 ceramic sample in air with different relative humidity: 10% (1), 34% (2), 55% (3), 82% (4). Data are presented as current density vs. electric field but the distribution of current over cross-section of a sample is inhomogeneous: current in peripheral regions is fairly high. Replotted using data from [24]

The high nonlinearity and high humidity sensitivity of studied SnO_2 varistor-sensor is attributed to the grain-boundary nature of conduction mechanism as it was stated earlier [21-25]. The grain-boundary potential barriers have appeared as a result of a sintering in air. Increase of applied voltage can lead to rather weak decrease in the barrier height at low voltages due to the trapping of electrons at the grain-boundary interface states followed by rather strong decrease in the barrier height at higher voltages (the physics of such voltage dependence of the barrier height in SnO_2 varistors can be similar to the situation in ZnO varistors [27]) . This assumption can explain observed nonlinear current-voltage characteristic at fixed relative humidity. At increase in relative humidity additionally adsorbed water molecules can dissociate at the surface according to the reaction $H_2O \leftrightarrow H^+ + OH^-$ and quite mobile proton H^+ can penetrate to the grain-boundary region. Due to the interaction of H^+ – ion with the chemisorbed oxygen O^-, the total negative charge trapped at the grain boundary becomes lower (on absolute value) leading to a decrease in the barrier height and subsequent increase in current at fixed voltage. The barrier mechanism of electrical conduction is supported independently by the observed increase of the capacitance of varistor-sensor in humid air (see Figure 6).

The effect of humidity sensitivity of low-field conductance in SnO_2 varistor ceramics is reversible and reproducible [21-25]. Therefore, this ceramic device is nonlinear and humidity sensitive. It can be used as the varistor-sensor – a device with double functions.

In Figure 4 the dependences of conductance on relative humidity are plotted for different SnO_2-based ceramics. Curve 1 in Figure 4 is for SnO_2-ZnO-CoO-SiO_2 material with relatively low nonlinearity coefficient $\beta \cong 10$ (at $j = 10^{-3}\ A \cdot cm^{-2}$) [21]. Curves 2 and 3 in Figure 4 are for SnO_2-based ceramics with the nonlinearity coefficient about $\beta \cong 30-50$ (at $j = 10^{-3}\ A \cdot cm^{-2}$) [23,24]. It is seen from Figure 4 that different SnO_2 –based ceramics show quite similar linear dependences of logarithm of the low-voltage conductance on the relative humidity.

According to the scanning electron microscopy (Figure 5), tin dioxide ceramic materials are not very dense. Therefore, pores play some important role in the penetration of humid air inside a sample. However, it seems that the mechanism of humidity influence on the conductance in SnO_2 –based ceramics is different from the humidity sensing mechanism in $MgCr_2O_4$-TiO_2 ceramics.

Sufficiently linear $\log\sigma - w$ dependence is observed in some ceramic materials where the conduction mechanism is rather different from the barrier-controlled [7]. Therefore, the only shape of $\log\sigma - w$ dependence cannot unambiguously define the mechanism of humidity sensitivity of conductance in oxide ceramics. But in the case if the conduction processes are controlled by the grain-boundary barriers, the linear $\log\sigma - w$ dependence is expected [23].

In [13] authors demonstrate linear response of logarithm of the resistance on relative humidity for single-crystalline Sb-doped SnO_2 whisker with Au contacts but in [12] linear dependence of the resistance (and not $\log R$) on relative humidity for single SnO_2 nanowire with Au electrodes (Au- SnO_2-Au) was found. The authors of [12,13] mention that current-voltage characteristics of Au- SnO_2-Au were linear but authors of [11] report nonlinear current-voltage characteristics of Au-SnO_2-Au. These contradictions can be caused by the structural and electrical differences between different SnO_2 materials and by the difference in the nature of the contact regions (assuming that there are no errors in the experiments in [11-13]). The situation can be quite complicated due to possible segregation of Sb in doped SnO_2:Sb nanomaterials [36]. These surface-segregated Sb atoms may form oxygen vacancy complexes at the surface and may be important active sites for oxygen and water adsorption [37]. In any case the existence of potential barriers at the contacts or inside the bulk even in the single-crystalline SnO_2 nanostructures cannot be excluded completely and, therefore, humidity can influence on the barriers which control the conduction.

In [16,17] the physical model of the grain boundary in ZnO varistor ceramics in the form of thin-film sandwich ZnO-Bi_2O_3 was obtained. Two-layer ZnO-Bi_2O_3 structure exhibit highly nonlinear current-voltage characteristic with strong current increase at about 4V [16] what is correlates with electrical properties of the single grain boundary in ZnO varistor [38]. The low-voltage current in this two-layer ZnO-Bi_2O_3 structure is quite strongly increased with the rise in relative humidity caused by the reduction of the potential barrier at the ZnO-Bi_2O_3 interface due to the decrease of the interfacial negative charge as a consequence of the adsorption of water molecules [17]. It was found as well that the increase in relative humidity causes the shift of the low-voltage part of current-voltage characteristic in ZnO varistor ceramics to higher currents though high-voltage highly nonlinear part of current-voltage characteristic is rather humidity insensitive [17]. It means that behavior of current-voltage characteristic in SnO_2 varistor ceramics [21,22] and in ZnO varistor ceramics [17] in air with different relative humidity is very similar. The variation in current-voltage characteristic in ZnO varistor ceramics in air with different relative humidity is similar to the observed one for two-layer ZnO-Bi_2O_3 structure [17]. Therefore, this similarity suggests that humidity sensitivity of conductivity in SnO_2 varistor ceramics [21-25] indeed is related to the reduction of the potential barriers at the grain boundaries.

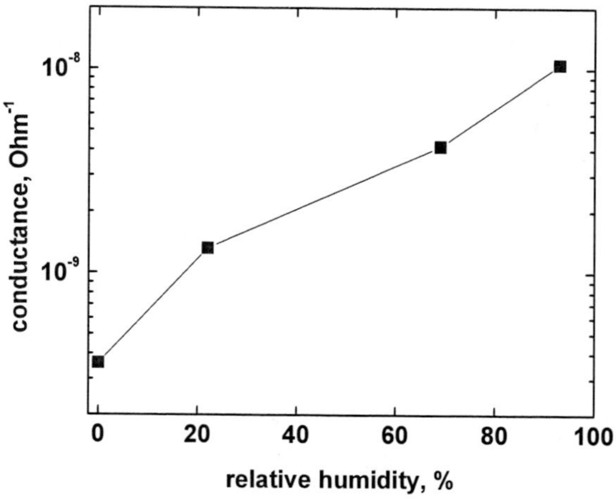

Figure 8. Dependence of conductance of ZnO-Bi_2O_3 two-layer structure on relative humidity. Replotted using data from [17]

In Figure 8 the relative humidity dependence of the conductance in ZnO-Bi_2O_3 two-layer structure is plotted. As it can be seen, the dependence $\log G - w$ is approximately linear. This fact explains why nearly linear dependence $\log \sigma - w$ was found for SnO_2 varistor ceramics (see Figure 4) where barrier mechanism of humidity sensitivity is expected and why the dependence of $\log G - w$ in the case of $MgCr_2O_4$-TiO_2 ceramics is rather nonlinear (see Figure 1).

4. Electrical Conduction in Semiconductor with Grain-Boundary Barriers

In this section the electrical conduction in oxide ceramics with electrically-active barriers at the grain boundaries is considered. These barriers determine the conductivity value.

4.1. Simplified Analysis of Conductivity versus Humidity Behavior

In the previous section it was shown that in SnO_2 ceramics, where electrical conduction is controlled by the grain-boundary barriers, experimental dependence of the low-field conductivity σ on the relative humidity w can be approximated by an empirical equation:

$$\sigma(w) = \sigma_0 \exp(Aw), \qquad (3)$$

where σ_0 is electrical conductivity in dry air and A is an empirical coefficient which characterizes the humidity sensitivity of conductivity.

The linear dependence of low-field conductivity on relative humidity in the scale $\log\sigma - w$ can be explained by the lowering of the barrier height in humid air [22,23]. The barriers in the sample are assumed to be identical. Two extreme situations can be considered: (i) humid air affects the whole sample homogeneously and (ii) humid air affects only peripheral part of the sample and the central part is left unchanged.

(i) It is assumed that in the first case the barrier height $\varphi(w)$ for all grain boundaries in the sample is decreased in humid atmosphere according to the linear relation:

$$\varphi\,(w) \cong \varphi_0 - \eta w, \tag{4}$$

where φ_0 is the barrier height at relative humidity $w = 0$, η is the constant. Then conductance G is increased with temperature according to the Arrhenius law:

$$G \propto \exp\left(-\frac{E_\sigma}{kT}\right), \tag{5}$$

where k is Boltzmann´s constant, T is the absolute temperature, E_σ is the activation energy (E_σ is about 1 eV): $E_\sigma = \zeta + \varphi_0$. Here ζ is the distance between the Fermi level and the conduction band edge inside the grain far from the boundary. It is assumed that the Fermi level in the bulk of tin dioxide grain is situated about several tenths of eV below the conduction band edge [39]. Therefore, $\zeta << \varphi_0$. It is expected that ζ does not depend on of relative humidity. Then conductance versus relative humidity can be expressed as:

$$G(w) \propto \exp\left(-\frac{\varphi\,(w)}{kT}\right) \propto \exp\left(-\frac{\varphi_0}{kT}\right)\exp\left(\frac{\eta w}{kT}\right). \tag{6}$$

From Eq.(6) it is seen that the linear relation between the logarithm of the conductance $\log G(w)$ and the value of relative humidity w takes place. From Figure 4 one can estimate the value of the coefficient η in Eq.(6) at T =300 K. For example, for SnO_2-Co_3O_4-Nb_2O_5-Cr_2O_3-Bi_2O_3 ceramics $\eta \cong 2.9\cdot10^{-3}\,eV(\%)^{-1}$ [23]. Therefore, the decrease in the barrier height at the variation of relative humidity between $w_1 = 10\%$ and $w_2 = 82\%$ is $\Delta\varphi_A = \eta\Delta w = \eta(w_2 - w_1) \cong 0.205eV$.

(ii) The decrease in the barrier height in humid air can be estimated as well in the case if humid air affects only peripheral part of the sample and the central part is left unchanged [23]. The lowering of the barrier height $\Delta\varphi_2$ is increased with relative humidity because more portion of current is flowing through the peripheral part of a sample. The calculated lowering of the barrier height $\Delta\varphi_2 \cong 0.217eV$ at relative humidity w = 82% and $\varphi = 1eV$ is slightly higher than the value $\Delta\varphi_1 \cong 0.205eV$ obtained above in the first case. For lower values of φ the higher values of $\Delta\varphi$ are found [23]: the lower φ values the higher current and higher $\Delta\varphi$ values in the peripheral parts of a sample are needed to reach certain current increase in humid air. In both cases the decrease in the barrier height about 20% takes place if the relative humidity is increased from 10 to 82%.

It would be interesting to obtain the expression for the approximation of conductivity-humidity characteristic $\sigma(w)$ which is related to the conduction process in a system with non-identical values of the barrier height and to find some parameter to characterize the sensitity of a humidity sensor. With this aim, let us consider the phenomenological model of humidity-dependent conduction of ceramic sensor with non-identical potential barriers at the grain boundaries. The basic idea is: the barrier height in oxide sensor material is decreased on relative humidity allowing respective current density increase.

Assume that the infinitely small increment of the current density Δj is proportional to the current density j and the increment of the relative humidity Δw :

$$\Delta j = \xi \cdot j \cdot \Delta w , \tag{7}$$

where ξ is dimensional coefficient. After an integration of Eq.(7) preliminary rewritten in differential form, the relationship between the current density and the relative humidity can be obtained:

$$j(w) = j_0 \exp(\xi w) , \tag{8}$$

where j_0 is an integration constant equal to the current density at zero relative humidity (in dry air). After the division of both parts of Eq.(8) on electric field E (Ohm's law takes place) the dependence of conductivity on relative humidity is obtained:

$$\sigma(w) = \sigma_0 \exp(\xi w) . \tag{9}$$

This equation is the same as Eq.(3) used for empirical approximation of experimental data. The weak point in this derivation is the assumption that the increment of the current density Δj is proportional to the current density j (see Eq.(7)). Below in the next section it is shown that this assumption is correct.

4.2. Electrical Conduction in Oxide Ceramics in Ohmic Region and in Dry Air

The sample of oxide sensor or varistor contains many grain boundaries (the thickness of a sample can be about 1 mm and cross-section can be about or less than 1 cm but the grain size in mentioned sensor or varistor ceramics can be about several micrometers). The grain boundaries in the same sample are rather different in their electrical parameters [26-29].

Assume that each grain boundary in the sample is characterized by the certain barrier height φ and φ-values for all grain boundaries in the sample are distributed homogeneously (for simplicity) from $\varphi_{\min}$ to $\varphi_{\max}$. If the barrier height scattering $\Delta\varphi = \varphi_{\max} - \varphi_{\min}$ in the sample is quite high, say, $\Delta\varphi \cong \varphi_{\max}$, the approach based on the percolation theory [40,41] can be applied.

Substitute the grain boundaries and grains in the sample by the resistor network. Replace each grain boundary with the barrier height φ by some resistor with resistance $R = R_0 \exp(\varphi / kT)$. Grains are assumed to be quite conductive, therefore, they are nodes in the resistor network. We have actually the Abrahams-Miller network [41].

Assume that initially all resistors are removed out of the sample. Then some resistors with $\varphi < \varphi_1$ (where $\varphi_{\min} < \varphi_1 < \varphi_{\max}$ and $\varphi_1 \cong \varphi_{\min}$) are imported into the sample and randomly occupied some vacant places. At this stage there is no conductance between the electrodes. There is no percolation in the system. Then φ_1 is slightly increased and new portion of resistors with $\varphi < \varphi_1$ is added into the sample. There is still no percolation in the system. This process is continued. At certain $\varphi_1 < \varphi_K$ imported resistors can belong to some finite cluster (group of resistors connected together but not connected to both electrodes at the same time). If to increase φ_1 gradually and discretely importing into the sample at each step some resistors with increasing φ and allow them to occupy randomly vacant places in the resistor network, then at $\varphi_1 = \varphi_K$ percolation in the system will appear: the last portion of imported resistors with $\varphi \cong \varphi_K$ will make electrical connection between the electrodes. The critical subnet of the infinite cluster (the group of resistors which have ensured percolation in the system at $\varphi \cong \varphi_K$) is created. At this step conductivity of the sample will jump from zero to $\sigma \propto \exp(-\varphi_K / kT)$ because resistors with $\varphi \cong \varphi_K$ give the main contribution to the conductivity of the sample. Resistors with lower barrier height φ have exponentially lower resistances and they are connected in series with critical resistors. If this process is continued, then resistors with exponentially higher resistances will be connected in parallel to the critical subnet of the infinite cluster but the value of conductivity of the sample $\sigma \propto \exp(-\varphi_K / kT)$ will not be changed significantly.

Grain-boundary barriers with the barrier height $\varphi \cong \varphi_K$ are called as "key barriers". Therefore, the conductivity of a sample is controlled by the key barriers with the barrier height φ_K ($\varphi_{\min} < \varphi_K < \varphi_{\max}$): $\sigma \propto \exp(-\varphi_K / kT)$. Key barriers in the sample are connected in series to barriers with lower barrier heights. Barriers with higher barrier heights than φ_K are connected in parallel to the critical subnet of the infinite cluster.

Therefore, the distance between key barriers L in general case can be greater or equal to the grain size l_g, $L \geq l_g$. In other words, if all grain boundaries are equal, then the average distance between them is equal to the grain size, $L = l_g$. If the variation of the barrier height $\Delta\varphi$ is high ($\Delta\varphi \cong \varphi_{\max}$), then $L > l_g$.

4.3. Electrical Conduction in Oxide Ceramics in Humid Air and in Ohmic Region

Let us consider now the situation if a sample with variable grain-boundary barriers is placed in air with relative humidity w. The procedure described above can be repeated and the barrier height $\varphi_K(w)$ of key barriers at relative humidity w can be found. Then the current density at relative humidity w can be written as:

$$j(w) = j_0 \exp\left[-\frac{\varphi_K(w)}{kT}\right]. \tag{10}$$

If the relative humidity is increased by infinitely small value Δw, then each barrier in the sample is decreased by some value, mentioned procedure can be repeated and, finally, the barrier height $\varphi_K(w+\Delta w) = \varphi_K(w) - \Delta\varphi_K$ of key barriers at relative humidity $w + \Delta w$ can be found. Here $\Delta\varphi_K > 0$ is the decrement of the barrier height. Then the current density at relative humidity $w + \Delta w$ can be written as:

$$j(w+\Delta w) = j_0 \exp\left[-\frac{\varphi_K(w) - \Delta\varphi_K}{kT}\right]. \tag{11}$$

It is assumed that due to a large number of grain-boundary barriers in the sample, the distribution of barriers on the barrier height is almost continuous. This assumption is rather rough but it helps in calculations. At infinitely small increment of the relative humidity Δw the increment of the current density

$$\Delta j(w) = j(w+\Delta w) - j(w) \tag{12}$$

can probably be infinitely small too. Then from Eqs.(10)-(12) we have:

$$\Delta j(w) = j(w)\left[\exp\left(\frac{\Delta\varphi_K}{kT}\right) - 1\right]. \tag{13}$$

If $\Delta\varphi_K << kT$, then infinitely small increment of the current density can be presented in a form:

$$\Delta j(w) \cong j(w)\frac{\Delta\varphi_K}{kT}. \tag{14}$$

The infinitely small decrement of the barrier height can be expressed through the infinitely small increment of the relative humidity Δw:

$$\Delta\varphi_K \cong \left(-\frac{d\varphi_K}{dw}\right)\Delta w, \tag{15}$$

where sign "-" takes into account that the barrier height of key barriers is decreased with relative humidity and the decrement of the barrier height $\Delta\varphi_K$ is positive (compare with Eq.(11)). Finally, the increment of the current density can be written from (14) and (15) by the next way:

$$\Delta j(w) \cong \xi \cdot j(w)\Delta w, \tag{16}$$

where

$$\xi = \frac{1}{kT}\left(-\frac{d\varphi_K}{dw}\right). \tag{17}$$

Humidity sensitivity coefficient ξ can be constant if the barrier height of key barriers φ_K is linearly decreased with relative humidity. Experimental data confirm that $\xi \cong const$ (see section 3.2.).

Eq.(16) coincides with Eq.(7). Therefore, the assumption (expressed in Eq.(7)) that the infinitely small increment of the current density Δj is proportional to the current density j and the increment of the relative humidity Δw is proven at least in the frames of this approach. It means that the dependence of the current density on the relative humidity is exponentially strong (see Eq.(8)). Indeed, changing in Eq.(16) from infinitely small increments to the differentials and integrating obtained differential equation, the dependence of the current density on the relative humidity is obtained in the form which coincides with Eq.(8). Respectively, conductivity versus relative humidity coincides with Eq.(9).

Experimental data obtained for SnO_2 varistor-sensor ceramics confirm that the dependence of electrical conductivity on relative humidity can be linearized in the scale $\log\sigma - w$ according to Eq.(9) (see Figure 4). It is interesting to note that in this case (where barrier nature of conduction looks convincing [21-25]) the curves in the scale $\log\sigma - w$ have only slightly increasing slope at higher relative humidity. On the contrary, in the case of $MgCr_2O_4$-TiO_2 ceramics where different conduction mechanism takes place, the dependence of electrical conductivity on relative humidity in the scale $\log\sigma - w$ clearly exhibits nonlinear behavior with decreasing slope at higher relative humidity [4] (Figure 1).

It is necessary to point out that linear dependence of the electrical conductance on the relative humidity in the scale $\log G - w$ was obtained for ZnO-Bi_2O_3 two-layer thin-film structure (Figure 8), where negative interface charge and potential barrier exist [17]. This fact confirms the model developed above. Mentioned ZnO-Bi_2O_3 two-layer thin-film structure exhibits nonlinear current-voltage characteristic and authors consider it as a model of the grain boundary in ZnO varistors [16,17]. C-U-measurements indicate that the negative charge at ZnO-Bi_2O_3 interface is decreased with humidity [17]. This reduces the potential barrier and leads to increase of current. Thus, humidity dependence of electrical conductance is explained by the reduction of the interfacial negative charge in humid air, though, the mechanism of such a reduction is not clear [17]. In any case, the results on ZnO-Bi_2O_3 two-layer thin-film structure [17] confirm our approach developed above for SnO_2 ceramic varistor-sensor.

4.4. Non-Ohmic Conduction in Oxide Ceramics in Dry Air

Consider now the dependence of electrical conductivity in semiconductor barrier system with non-identical grain-boundary barriers (which control conduction) on electric field in dry air. With this purpose it is necessary to repeat two times in the non-Ohmic regime the procedure similar to key barriers finding in the Ohmic regime used above for conduction in humid air (see section 4.3.). It is useful to stress that repeating this procedure can lead, in principle, to the different percolation paths at each electric field. Assume that the condition of an applicability of the percolation theory (wide distribution of the grain-boundary barrier heights in a sample) is fulfilled and respective values $\varphi(E)$ and $\varphi(E+\Delta E)$ are found.

The argumentation remains valid if substitute the relative humidity w for the electric field E (see Eq.(14)). Then infinitely small increment of the current density at infinitely small increment of the electric field ΔE can be presented in a form:

$$\Delta j(E) \cong j(E)\frac{\Delta\varphi_K}{kT}, \tag{18}$$

where the current density through a sample of ceramic varistor (but not only the current density through a single grain boundary) is:

$$j = j_0 \exp(-\frac{\varphi_K(E)}{kT}). \tag{19}$$

Here $\varphi_K(E)$ is the barrier height at electric field E. In spite of the sample contains barriers with quite different barrier heights from $\varphi_{\min}$ to $\varphi_{\max}$, some value $\varphi_K(E)$ (which is between $\varphi_{\min}$ and $\varphi_{\max}$) determines the conductivity of the sample (see Eq.(19)). The infinitely small decrement of the barrier height in Eq.(18) can be expressed through the infinitely small increment of the electric field ΔE :

$$\Delta\varphi_K \cong \left(-\frac{d\varphi_K}{dE}\right)\Delta E . \tag{20}$$

Sign “minus“ takes into account that the barrier height of key barriers is decreased with the electric field. The decrement of the barrier height $\Delta\varphi_K$ is positive. Finally, the increment of the current density can be written as:

$$\Delta j(E) \cong \alpha \cdot j(E)\Delta E , \tag{21}$$

Where

$$\alpha = \frac{1}{kT}\left(-\frac{d\varphi_K}{dE}\right). \tag{22}$$

Eq.(21) gives the increment of the current density related to non-Ohmic behavior. To obtain complete increment of the current density, it is necessary to add the increment of the current density related to Ohmic behavior $\Delta j_{Ohm} = \sigma(E)\Delta E = (j / E)\Delta E$. Here $\sigma(E)$ is the conductivity of a material at the electric field E. Then finally the differential equation can be obtained in such form:

$$dj(E) = (j / E)dE + \alpha \cdot j(E)dE . \tag{23}$$

After the integration the dependence of the current density on the electric field $j = \sigma_0 E \exp(\alpha E)$ is obtained (see Eq.(2)). Here σ_0 is a constant of integration having the meaning of the conductivity at low electric field (Ohmic conductivity) and α is the nonlinearity factor (see Eq.(22)).

Earlier Eq.(2) was obtained by the similar way but Eq.(21) was simply assumed [26-28]. The principal difference of this work is that Eq.(21) now is obtained by calculation. This result confirms earlier made assumption.

Obtained here equation $j = \sigma_0 E \exp(\alpha E)$ (Eq.(2)) is quite similar with the earlier obtained expression $j = \sigma_0 E \exp(BE^{1/\nu})$ (where $\nu = 0.9$ is the critical index of the correlation length in percolation theory) for the field dependence of the current density in disordered polycrystalline semiconductor in the assumption of a filling of grain-boundary states under applied voltage [42] and it is substantially different from the expression $j = \sigma_0 E \exp(DE^{1/(1+\nu)})$ [40] found in the assumption that grain-boundary states are already filled before the application of voltage. In ZnO varistors [27,43] and in SnO_2 varistors [25,44] grain-boundary states are filling under applied voltage (in the range of low and medium voltages). Therefore, Eq.(2) can be used for the approximation of current-voltage characteristics of ZnO and SnO_2 varistors what was demonstrated in our previous publications [26-28].

4.5. Non-Ohmic Conduction in Oxide Ceramics in Humid Air

Let us consider the dependence of electrical conductivity in the barrier system on electric field in humid air. It would be interesting to understand why the nonlinearity factor α (the slope of $\ln\sigma - E$ dependence) at higher humidity becomes lower (see Figure 3). It is necessary to make an assumption about the interaction of humid air with a sample. Let us consider two cases: (i) each grain boundary in a sample is in air with the same relative humidity and (ii) humid air penetrates only into the surface (peripheral) area of a sample.

(i) Assume first that humid air penetrates to all grain boundaries. In this case the barrier height at each value of the electric field becomes lower than in dry air and the slope of $\varphi_K = \varphi_K(E)$ function in humid air can be lower than in dry air. Therefore, it is assumed that at certain increment of the electric field in the sample in humid air the respective decrement of the barrier height is lower than in the same sample in dry air and the derivative of the barrier height on the electric field in Eq.(20) (the slope of $\varphi_K = \varphi_K(E)$ function) is decreased and this decrease is linearly depends on the relative humidity w. Then Eq.(20) can be modified by the next way:

$$\Delta\varphi_K \cong \left[\left(-\frac{d\varphi_K}{dE}\right)_{w=0} - \chi w\right]\Delta E, \tag{24}$$

where index $w = 0$ means "at zero relative humidity " ("in dry air"), χ is the dimensional coefficient. Then, similarly, the dependence of the current density on the electric field is obtained:

$$j = \sigma_1 E \exp(\alpha_w E), \tag{25}$$

where σ_1 is the electrical conductivity at low electric field in humid air and the nonlinearity factor in humid air is:

$$\alpha_w = (kT)^{-1}\left[\left(-\frac{d\varphi_K}{dE}\right)_{w=0} - \chi w\right] = \alpha - \Delta\alpha . \tag{26}$$

Here $\Delta\alpha = (kT)^{-1}\chi w > 0$. Therefore, the nonlinearity factor in humid air is decreased according to the experiment (Figure 3).

(ii) If humid air penetrates only into the surface (peripheral) areas of a sample, then these parts become more conductive than the bulk [21,23]. Therefore, the conductance at high electric field and at high humidity can be approximately presented as a sum of non-Ohmic conductance in dry air (first term in the left part) and conductance induced by high humidity (second term in the left part):

$$\sigma_0 e^{\alpha E} + \sigma_w \cong \sigma_1 e^{\alpha_w E} . \tag{27}$$

The right part is the resulting non-Ohmic conductance in humid air (see Eq.(25)). For simplicity, the transition between "conductivity" and "conductance" are taken into account in the pre-exponential factors in Eq.(27). At zero electric field ($E = 0$) from Eq.(27) we have the equation:

$$\sigma_0 + \sigma_w = \sigma_1 , \tag{28}$$

which expresses that low-field conductance in humid air is a sum of ohmic conductance in dry air and conductance induced by humidity. Assume that the electric field is fixed, $E = E_0$, where E_0 is selected in non-Ohmic region. If $\sigma_w << \sigma_0$, then humidity practically has no influence on the conductance. In the experiment rather the case $\sigma_w > \sigma_0$ or even $\sigma_w >> \sigma_0$ takes place. Considering the last condition, from Eqs. (27) and (28) we have:

$$\alpha_w \cong E_0^{-1} \ln(1 + \sigma_0 \sigma_w^{-1} e^{\alpha E_0}) . \tag{29}$$

If the nonlinearity factor in humid air is not equal to zero (σ_w is not too high), then the second term in (29) dominates and, finally, we have $\alpha_W \cong \alpha - \Delta\alpha$, where $\Delta\alpha \cong E_0^{-1} \ln(\sigma_w \sigma_0^{-1}) > 0$. Therefore, in both cases the estimations give lower nonlinearity factor α (the slope of the dependence $\ln \sigma - E$) at higher humidity what is in accordance to experimental data (see Figure 3).

It is necessary to stress that mentioned above "conductance induced by humidity" is rather due to the lowering of grain-boundary potential barriers in humid air than appearance of conductive path, for example, by mobile protons or other reasons (see [7]) because the increase in capacitance takes place [23,30] and even surface part of a sample exhibits highly nonlinear current-voltage characteristic [23].

Therefore, the dependence of conductivity on electric field becomes less strong at higher relative humidity (see Figure 3). The dependence of conductivity on relative humidity at higher electric fields becomes even saturated [21-25] due to a prevalence of the bulk non-Ohmic conductance which is insensitive to the variations in relative humidity. Practically unchangeable highly nonlinear current-voltage characteristic in SnO_2-based ceramics was observed in a wide range of relative humidity $w = 10 - 90\%$ [22,24].

CONCLUSION

Experimental dependences of electrical conductivity on relative humidity in $MgCr_2O_4$-TiO_2 ceramics (where conduction mechanism in humid air is rather related to porous structure and direct participation of mobile protons) and in SnO_2-based varistor ceramics (where conduction mechanism in humid air is determined by electron transport across grain-boundary barriers with the barrier height dependent on relative humidity) are slightly different. The logarithm of conductivity versus relative humidity is nonlinear ($MgCr_2O_4$-TiO_2 ceramics) and linear (SnO_2-based ceramics). The functional dependences of electrical conductivity on the relative humidity and on the electric field are calculated. It is shown that electrical conductivity exponentially strongly depends on the relative humidity and on the electric field. Calculated characteristics are in an agreement with the experimental data for SnO_2-based varistor-sensor ceramics.

ACKNOWLEDGMENTS

The author would like to thank Dr. R. Bulpett and G. Jimenez-Santana for their assistance with the scanning electron microscopy.

REFERENCES

[1] Nitta, T. *Ind Eng Chem Prod Res Dev*., 1981, 20, 669-674.
[2] Seiyama, T; Yamazoe, N; Arai, H. *Sens Actuat B*, 1983, 4, 85-96.
[3] Kulwicki, BM. *J Phys Chem Solids*., 1984, 74, 1015-1031.
[4] Nitta, T. In *Chemical Sensor Technology*; T. Seiyama, Ed; Kodansha Ltd.: Tokyo / Elsevier: Amsterdam-Oxford-New York-Tokyo, 1988, Vol.1, 57-78.
[5] Kulwicki, BM. *J Am Ceram Soc*., 1991, 74, 697-708.
[6] Pelino, M; Cantalini, C; Faccio, M. *Act Pass Elec Comp*., 1994, 16, 69-87.
[7] Traversa, E. *Sens. Actuat B*, 1995, 23, 135-156.
[8] Chen, Z; Lu, C. *Sens Lett*, 2005, 3, 274-295.
[9] Shimizu, Y; Arai, H; Seiyama, T. *Sens. Actuat*, 1985, 7, 11-22.
[10] Batzill, M; Diebold. U. *Prog Surf Sci*., 2005, 79, 47-154.
[11] Li, D; Wang, Y; Xia, Y. *Nano Lett*, 2003, 3, 1167-1171.
[12] Kuang, Q; Lao, C; Wang, Z.L; Xie, Z; Zheng. L. *J Am Chem Soc*., 2007, 129, 6070-6071.
[13] Huang, J; Wang, J; Zhukova, AA; Rumyantseva, MN; Gaskov, AM; Yu, K; Gu, C; Liu, J. *Sens Lett*, 2009, 7, 1025-1029.
[14] Chatman, S; Poduska, KM. *ACS Appl Mater Interf*. 2009, 1, 552-558.
[15] Shimizu, Y; Di Bartolomeo, E; Traversa, E; Gusmano G; Hyodo, T; Wada, K; Egashira, M. *Sens. Actuat B*, 1999, 60, 118-124.
[16] Suzuoki, Y; Ohki, A; Mizutani, T; Ieda. M. *J Phys D Appl Phys*., 1987, 20, 511-517.
[17] Suzuoki, Y; Ohki, A; Mizutani, T; Ieda. M. *J Phys D Appl Phys*., 1987, 20, 518-521.
[18] Glot, A; DiBartolomeo, E; Traversa, E. *J Eur Ceram Soc*., 1999, 19, 715-719.

[19] Glot, AB; Chack, AM; Chernyi, BK; Yakunin, AYa. *Inorg Mater*, 1974, 10, 1866-1868.

[20] Yakunin, AYa; Chernyi, BK; Chack, AM; Glot, AB. *Inorg Mater*, 1976, 12, 803-805.

[21] Glot, AB. *Inorg Mater*, 1984, 20, 1522-1523.

[22] Skuratovsky, I; Glot, A; Di Bartolomeo, E; Traversa E; Polini, R. *J Eur Ceram Soc.*, 2004, 24, 2597-2604.

[23] Skuratovsky, I; Glot, A; Traversa E. *Mater Sci Eng B*, 2006, 128, 130-137.

[24] Gaponov, AV; Glot, AB; Ivon, AI; Chack, AM; Jimenez-Santana, G. *Mater Sci Eng B*, 2007, 145, 76-84.

[25] Glot, AB; Sandoval-García, AP; Gaponov, AV; Bulpett, R; Jones, BJ; Jimenez-Santana, G. *Adv Tech Mat Mat Proc J*, 2008, 10, 21-32.

[26] Glot, AB. *J Mater Sci.*, 2006, 41, 5709-5711.

[27] Glot, AB. *J Mater Sci Mater Electron*, 2006, 17, 755-765.

[28] Glot, AB; Skuratovsky, IA. *Mater Chem Phys.*, 2006, 99, 487-493.

[29] Gaponov, AV; Glot, AB. *J Mater Sci Mater Electron*, 2010, 21, 331-337.

[30] Glot, AB; Bulpett, R; Nadzhafzade, AM; Skuratovsky, IA. *Photoelectronics*, 2001, No.10, 47-49.

[31] Glot, AB; Zlobin, AP. *Inorg Mater*, 1989, 25, 274-276.

[32] Pianaro, SA; Bueno, PR; Longo, E; Varela, JA. *J Mater Sci Lett*, 1995, 14, 692-694.

[33] Bueno, PR; Varela, JA; Longo, E. *J Eur Ceram Soc.*, 2008, 28, 505-529.

[34] Metz, R; Morel, J; Houabes, M; Pansiot, J; Hassanzadeh, M. *J Mater Sci.*, 2007, 42, 10284-10287.

[35] Parra, R; Ponce, MA; Aldao, CM; Castro, MS. *J Eur Ceram Soc.*, 2007, 27, 3907-3909.

[36] Fliegel, W; Behr, G; Werner, J; Krabbes, G. *Sens. Actuat B*, 1994, 18-19, 474-477.

[37] Dusastre, V; Williams, DE. *J Phys Chem B*, 1998, 102, 6732-6736.

[38] Einzinger, R. *Ber Dt Keram Ges*, 1975, 52, 244-246.

[39] Jarzebski, ZM; Marton, JP. *J Electrochem Soc.*, 1976, 123, 299C-310C.

[40] Shklovskii, BI. *Sov Phys Semicond*, 1979, 13, 93-97.

[41] Shklovskii, BI; Efros, AL. Electronic properties of doped semiconductors. *Solid State Sci.,* 45, Springer, 1984.

[42] Vinnikov, AYa; Meshkov, AM; Savushkin, VN. *Solid State Phys.*, 1982, 24, 1352-1359.

[43] Mahan, GD; Levinson, LM; Philipp, HR. *J Appl Phys.*, 1979, 50, 2799-2812.

[44] Glot, AB; Gaponov, AV; Sandoval-García, AP. *Phys B Cond Matter*, 2010, 405, 705-711.

In: Relative Humidity: Sensors, Management...
Editor: John M. Castillo

ISBN: 978-1-61761-734-8

Chapter 4

OPTICS OF AIRBORNE PARTICLES UNDER ELEVATED HUMIDITY CONDITIONS

Miroslav Kocifaj*

ICA, Slovak Academy of Sciences,
Dúbravská road 9, 845 03 Bratislava, Slovak Republic.

ABSTRACT

The amount of radiative energy reaching the Earth's surface depends on many factors. Except for geometrical relations (such as a distance Earth-Sun, elevation of the Sun, and orography of the terrain surrounding a measuring site), the direct solar radiation as well as the diffuse radiation undergo complex alteration with regard to the actual meteorological conditions and the physical state of the atmosphere. Under clear sky conditions, the attenuation of the visible radiation is mainly due to scattering and absorption by air molecules and aerosol particles. These processes are linked, but, especially aerosols are responsible for rapid changes of the radiative fluxes. The effect of aerosols on the solar radiation becomes more important under elevated relative humidity (RH) conditions. The size of aerosol particles grows significantly when RH > 75% and the solid particles made up of the deliquescent materials may transform to solution droplets. Thus during the humidification process the optical properties of aerosol particles evolve. Definitely, the optical responses of humid particles will differ from those of non-humid particles. Therefore the RH is one of the crucial factors affecting the availability of radiative energy at the ground, implying a necessity to incorporate RH into optical models of the atmosphere. In this chapter the optical effects by single aerosol particles and particle polydispersions are discussed and some numerical demonstrations are shown.

Keywords: relative humidity – aerosol particles – solar radiation – diffuse radiation – atmospheric optics.

* Corresponding author: E-mail: kocifaj@savba.sk

1. Introduction

Solar radiation penetrating the Earth's atmosphere undergoes many changes due to absorption and scattering by air molecules and aerosol particles. The air molecules scatter the sunbeams in accordance with the Rayleigh theory (Bohren and Huffman, 2004), for which the intensity of scattered beams is inversely proportional to the fourth power of wavelength. In addition, the atmospheric gases (such as ozone, oxygen, or water vapor) are responsible for selective absorption at various wavelengths in the visible spectral range (Krueger and Mizner, 1976; Kato et al. 1997, Golovko, 2000). Both effects, scattering and absorption, influence together the reduction of the flux density of the incident radiation. The direct solar radiation is partly absorbed and partly transformed to diffuse radiation. The sum of scattering and absorption is commonly called extinction. The spatial and temporal variations of the extinction due to atmospheric gases are much smaller than that caused by aerosol particles.

The particulate matter is one of most optically unstable constituents of the atmosphere (Hansen et al., 1997). It is well recognized that the characteristics of the scattered radiation field strongly depend on the size distribution, spatial concentration, and chemistry of the aerosol particles. All these aerosol properties are changing in transforming air masses (Esposito et al., 2004) and the aerosol loading can be especially important in densely polluted and industrialized regions (Latha and Badarinath, 2005). Aerosols are usually originated from sources of pollution surrounding the measuring places (Bukowiecki et al., 2002), but the particles may also be transported by wind to long distances (Pickering et al., 1996). Note that a detail analysis of backward trajectories is usually needed to associate the aerosol polydisperse system with the source of pollution (Niemi et al., 2004).

Freshly emitted aerosol particles may rapidly grow under specific circumstances. The soot particles may collide and stick together in later stages of combustion processes (Lahaye and Prado, 1981; Charalampopoulos, 1992). More typically, the small solid particles made up of deliquescent materials become solution droplets at high humidity. As the sizes of particles grow in the process of humidification, their shapes as well as bulk refractive indices will evolve too. In particular, irregularly shaped grains may take the spherical forms because of high water content. Interaction of the solar radiation with the particles strongly depends on their topology. Specifically, the optical response of the multi-component particle quickly changes with internal re-distribution of individual substances. For instance, absorption by the carbonaceous aerosol particle is much smaller if the carbon species are centered in the core rather then being distributed at the edges of the particle. In the first case the beams can penetrate through the shell, while in the second case most of these beams are blocked by optically dense carbon segments. In general, the different topologies mean the different scattering patterns. A slight change in the interfaces between any two material constituents may translate to a measurable change in the optical characteristics of scattered light. This refers to the particles built of slightly and strongly absorbing elements. One of the typical examples is an aggregated particle composed of inorganic ions, carbonaceous material, crustal elements, and water (Moya et al., 2004). Strongly absorbing carbonaceous materials are traditionally originated from industrial combustion or biomass burning. In urban environments the black carbon species are frequently below 10% of the particle volume content (Horvath et al. 1996). But, in densely polluted industrial areas the carbonaceous species may sporadically contribute 20–50% to the fine particulate mass (Park et al., 2001).

The rest material mixtures are usually slightly absorbing. The bulk optical properties of a composite particle must be then simulated numerically using information on particle topology and volume content of individual constituents.

2. Aerosol Particles Under Elevated Humidity Conditions

The simplest way for determining the bulk refractive index of a mixed-phase aerosol particle is to calculate the weighted contribution of all material constituents. Translated to the mathematical formula, the mean complex refractive index of the particle is $m = \sum_j m_j f_j$, where m_j is the refractive index of j-th component and f_j is its volume fraction. Although the above approach appears to be well understandable, such a volume-averaged mixing rule has no theoretical foundation. More frequently the Bruggeman (1935) mixing rule is applied. The Bruggeman theory assumes the inhomogeneous material to have an aggregate structure, a random mixture of two or more materials (Chýlek and Videen, 1998; Yang el at., 2000).

The material composition of solid-phase aerosols depends to a great extent on their origin and it may quickly change as the particles evolve in the atmospheric environment. Real part of the refractive index of atmospheric aerosols usually ranges from 1.3 to 1.65. The small values of 1.35 are known for e.g. maritime aerosols, while larger values can be typical for Saharan dust (Borde and Verdebout, 2003). The imaginary part m_i of the refractive index varies in a wider range (over several orders of magnitude), from 10^{-9} to more than 0.1 (Sokolik and Toon, 1999). The relatively small values of 10^{-5}-10^{-2} are representative for silicates (Volten et al., 2005), while urban or soot aerosols are quite absorbing with the imaginary part of the refractive index larger than 0.1.

The numerical simulations in this chapter were performed for urban/rural-like aerosols. Real part of the refractive index is assumed to be $m_r \approx 1.55$ showing a slight dependence on the wavelengths in the visible spectral range. Rural and urban aerosols often consist of a mixture of scattering and absorbing materials, so Kent et al. (1983) accepted the m_i to be about 0.1. As the relative humidity grows the water condensation results in a continuous coating of the solid dust core. In this process the bulk refractive index approaches that of water. Mutual changes of real and imaginary parts of the complex refractive index are presented in Figure 1, where the state of humidification is indicted by percentage values.

The solid particle has notoriously a stochastic morphology (Figure. 2), thus the water condensation is a complex process. In principle, the waterish coating can be discontinuous at the early phases of water condensation. In practically important case it means that nucleus-air and nucleus-water interfaces may coexist when the water content is low. This can have certain effects on the interaction of electromagnetic radiation with such a particle. It is well-known that penetration depth for light-beams generally is increased if the media on either side of the interface have similar refractive indices. The air-nucleus interface tends to reflect more light than the water-nucleus interface. Therefore the optical behavior of a composite particle will definitely vary with its topology.

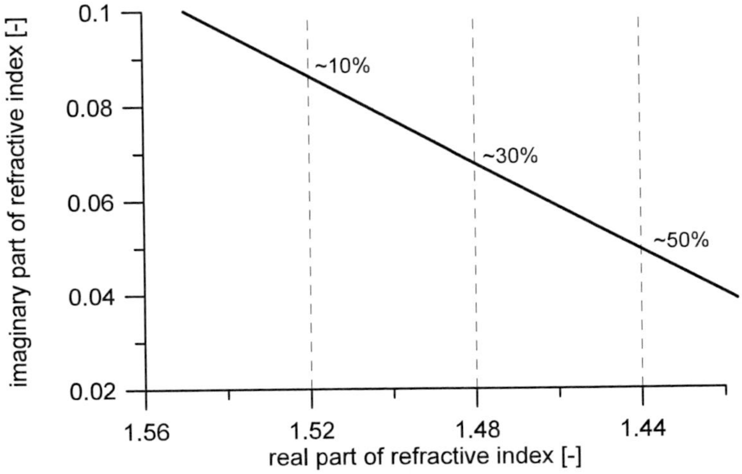

Figure. 1. Refractive index of urban aerosols evolving in the process of humidification. The water content of the aerosol particles is indicated by percentage values.

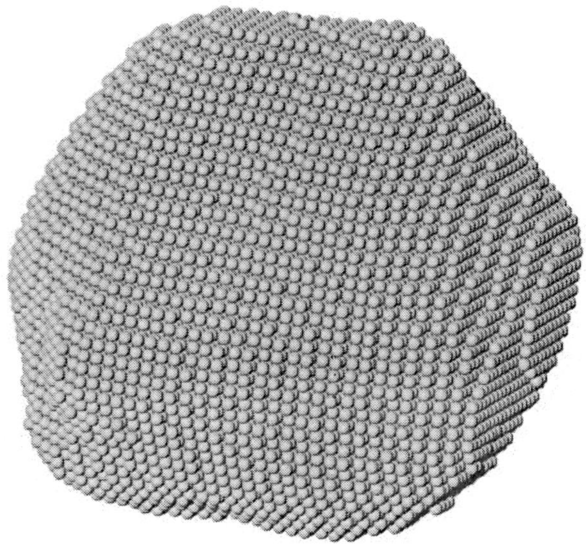

Figure. 2. Model of perturbed sphere – a Gaussian solid-phase aerosol particle.

As the water content of an aerosol particle increases, the particle shape becomes more spherical. In the same time, the overall size of the particle grows too. It is therefore obvious that a process of humidification is followed by many changes in particle microphysical characteristics. Specifically, size, shape, topology and refractive index of a particle alter one depending on the others. These changes are normally related to the original microphysical properties of non-humid nucleus. For instance, the rate of particle growth is a function of original particle size (Hänel, 1976; Shettle and Fenn, 1979).

3. Optical Behavior of Single Particles

The radiation field in an environment containing the particulate matter is formulated in terms of the optical properties of particle ensembles with defined size or shape distributions. Every single particle scatters the incident radiation thus generating the diffuse radiation field. In addition, some part of the radiation is absorbed increasing the internal energy of the particle. This amount of energy is usually consumed for heating the particle. Both, scattering and absorption processes are responsible for the attenuation of the incident electromagnetic energy. A common effect of all particles can be then easily measured by photometers. A correct interpretation of experimentally determined data is only possible if the basic optical behavior of a single particle is known. In other words, the optics of aerosol particles under evolving water contents has to be analyzed.

Many studies on aerosol optics are still based on the conventional Mie theory (Mie, 1908) exclusively applicable to spherical and homogeneous particles. Although the shape of a humid particle approaches that of a sphere, the internal distribution of material constituents is neither homogeneous, nor regular. It is therefore obvious that the optical response of a real particle can differ from that of a homogeneous sphere.

The optical response of a scattering particle interacting with electromagnetic radiation may quickly change if the wavelength λ of the incident radiation is of the same order as the size of the particle. Especially, the backscattered signals are quite sensitive to small variations in particle sizes. It is not difficult to show to what extent the scattering features of homogeneous spheres and inhomogeneous irregularly shaped particles differ. However, the situation is more complex if these particles act as the condensation nuclei in a process of humidification. The outer shape of a humid particle approaches that of a sphere under elevated humidity conditions, and the possible deviations in optical responses are related to the internal interfaces between water coating and the solid nucleus. During early phases of humidification, the transition from the dry to the humid phase may be important, because of different water coverage formation. It can be assumed that a water coat on an ideally spherical particle will show a spherical symmetry, because of no position-specific preferences. In contrary, the water shell on a non-spherical nucleus might not evolve in the same way. It can be assumed that any cavity on the particle surface will be filled as the first, so the thickness of the water coverage will depend on particle morphology. An easiest way how to simulate the water coat on a non-spherical particle is a continuous filling of all holes on the particle's surface, assuming that the water shell has the tendency to keep a spherical form. It means that water-core, water-air and air-core interfaces may co-exist in early phases of humidification. The air-core interface diminishes as the water content increases. Under elevated humidity the air-core interface definitely disappears. Such a mechanism may not exactly satisfy the processes for many particles. However, we anticipate that the errors in the optical properties of mixed-phase aerosols calculated this way are relatively small (Kocifaj et al., 2008).

We have built a model of the perturbed sphere (Figure. 2) and applied it in numerical simulations. The motivation for using such slight surface deformations was to analyze minimum conditions under which the optical properties of the particle become deviating from those of homogeneous spheres. Simultaneously, the comparative study on two component concentric sphere is performed. Here we consider a compact core coated by a water shell, both do not interacting together. In that case the interface between core and shell keeps a perfectly spherical form. The electromagnetic problem for the coated sphere can be solved analytically (Wickramasinghe, 1963), following the concept of the Mie theory. At present, there are several powerful numerical implementations applicable to calculate bulk optical properties of core-mantle particles that are concentrically stratified spheres (consult the light scattering software freely available at http://www.t-matrix.de). This kind of a particle model allows us to use the separation variable concept (Kahnert, 2003), in which the scalar Helmholtz equations are solved. The analytical solution can be translated to extremely fast and accurate numerical schema (e.g. C-M SPH ver. 3.3i by Farafonov et al., 1996). Unfortunately, no analytical solution exists for a non-spherical particle, thus the computations on perturbed sphere coated by a water shell were done by means of Discrete Dipole Approximation (DDA) (Draine, 1988). Although the T-matrix method appears to be more convenient for this kind of morphologies, we have used DDA due to the following facts: In the last years we have built a large database of the optical properties of irregular non-homogeneous particles. DDA is one of well-suited tools for modeling such particles. To assemble the calculated data sets in a common format, and to make the results easily comparable, the DDA is preferred as a standard tool for these purposes. However, the DDA places a strong requirement on the size of particle's dipole model. Specifically, the distance between individual cells (dipoles) must be small enough to satisfy the condition that a product of $|m|k\,d$ is smaller than unity. Here m is the complex refractive index, k is the wavenumber, and d is lattice spacing, i.e. the characteristic size of a cell. The numerical results obtained for homogeneous sphere, core-mantle sphere, and Gaussian stochastic core with varying water coverage are presented and discussed in Figures. 3-9.

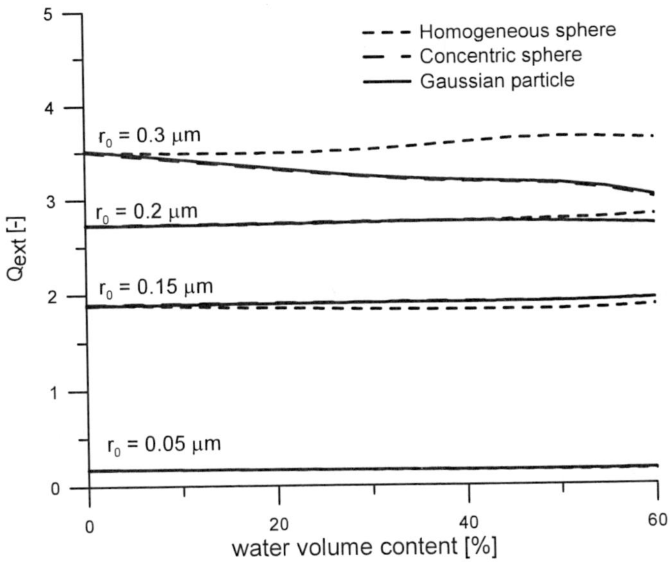

Figure. 3. The evolution of efficiency factors for extinction during humidification of urban aerosol particles. Both, the topology model as well as the original radius r_0 of the particles are indicated in the figure. Refractive index of solid nucleus is $1.55{+}0.1i$ (in accordance with Kent et al., 1983).

The attenuation of incident radiation is frequently expressed in term of efficiency factor for extinction Q_{ext}. This factor is linearly proportional to the extinction cross section, thus it characterizes an amount of energy removed by a particle when placed into electromagnetic field. The larger Q_{ext}, the larger attenuation of electromagnetic waves. The results of numerical simulations for all three particle models are depicted in Figure. 3. Initial radius r_0 of a non-humid particle is indicated in the figure. A general trend is that Q_{ext} for the particles with Gaussian nuclei and the core-mantle spheres are almost identical for any water content. On the one hand, the extinction efficiencies for all particle models coincide well at the early phases of humidification. On the other hand, a homogeneous sphere and a coated particle show different behavior of Q_{ext} under elevated humidity. The discrepancies between a homogeneous sphere and other particles are even more important when dealing with efficiency factor for scattering Q_{sca} (Figure. 4). It is obvious, that increased extinction efficiency for a homogeneous sphere with an initial radius 0.3 μm is caused by its enhanced scattering efficiency. This finding is also supported by calculated asymmetry parameter for scattering $\langle \cos \rangle$ (Figure. 6). Note that asymmetry parameter is the cosine-weighted integral of the aerosol-scattering phase function. In other words, it is a measure of the electromagnetic energy that is scattered to the forward direction. If only forward scattering realizes, the $\langle \cos \rangle$ =1. If $\langle \cos \rangle$=-1 the scattered radiation is exclusively directed toward backward. For isotropic scattering we obtain $\langle \cos \rangle$=0. Typically the values of asymmetry parameter lie between 0.5 and 0.95 for many aerosol particles.

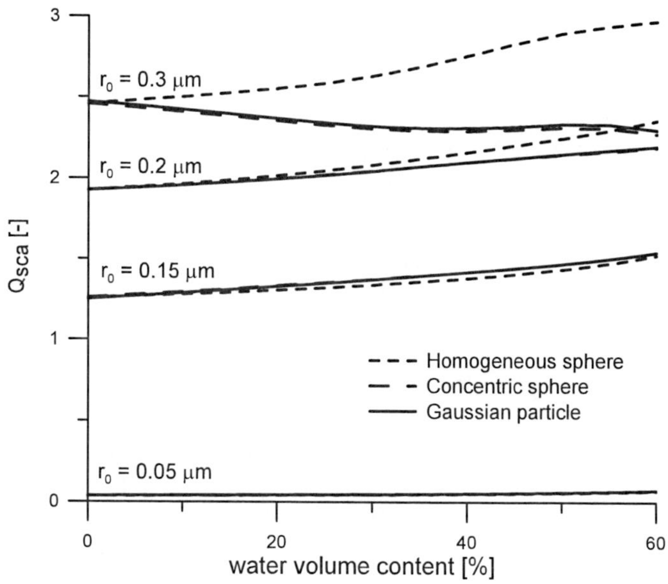

Figure. 4. The same as Figure. 3 but for scattering efficiency factors.

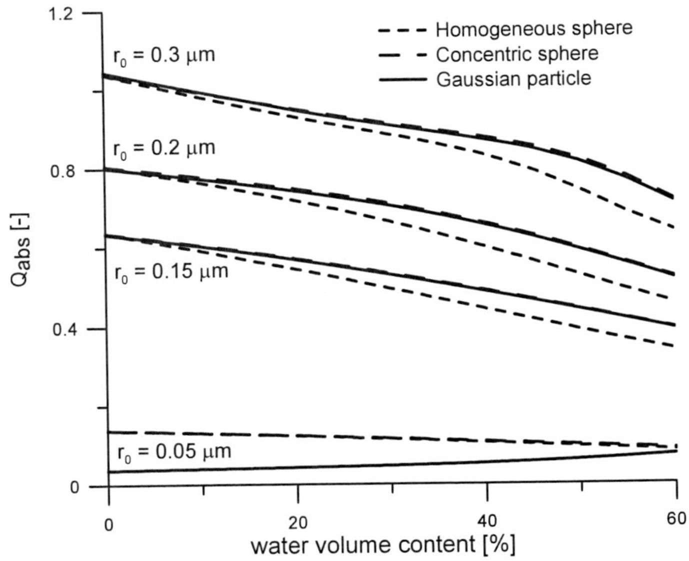

Figure. 5. The same as Figure. 3 but for absorption efficiency factors.

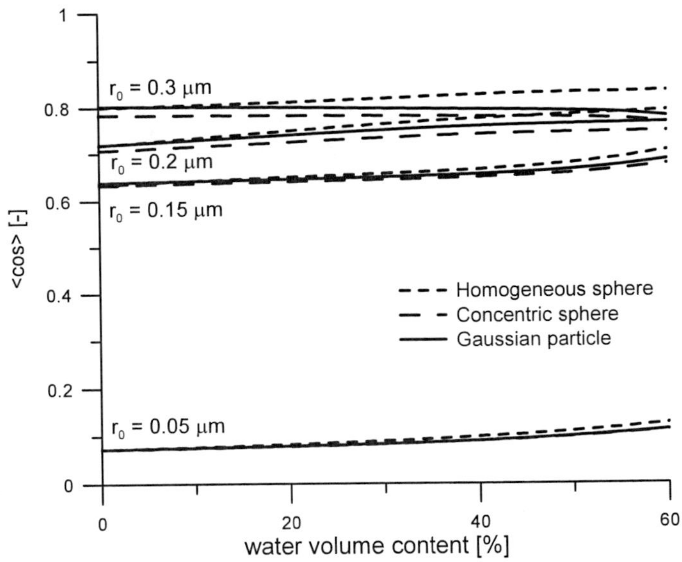

Figure. 6. The same as Figure. 3 but for asymmetry parameter.

It is shown in Figures. 4 and 5 that the idealized assumption for homogeneous sphere overestimates the scattering efficiency and underestimates the absorption efficiency, both under elevated humidity. The attenuation of electromagnetic energy seems to be overestimated if the initial radius of homogeneous sphere is above 0.2 μm. However, this can change if the particle radii are larger than the wavelength of an incident radiation.

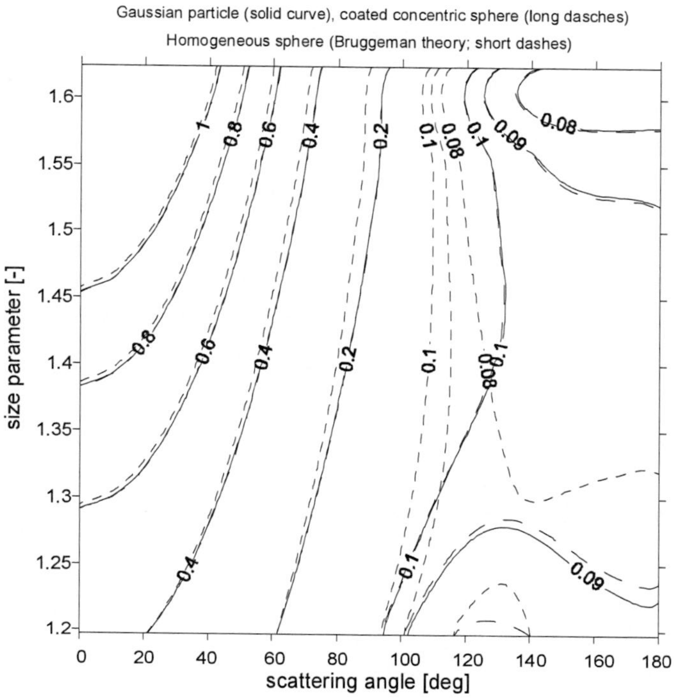

Figure. 7. Scattering phase function S_{11} for humid aerosol particles. The growth of the size parameter $x = 2\pi r_0 / \lambda$ reflects a process of humidification for the particles with original volume-equivalent radii $r_0 = 0.1$ μm. Refractive index of solid material is 1.55+0.1i (in accordance with Kent et al., 1983).

A bit lower scattering efficiency of a homogeneous sphere with initial effective radius $r_0 < 0.15$ μm is proven by angular behavior of scattering phase function $S_{11}(\theta)$ (Figure. 7). The phase function $S_{11}(\theta)$ is the element of so-called scattering matrix (van de Hulst, 1957)

$$\begin{pmatrix} S_{11} & S_{12} & 0 & 0 \\ S_{12} & S_{11} & 0 & 0 \\ 0 & 0 & S_{33} & S_{34} \\ 0 & 0 & S_{34} & S_{33} \end{pmatrix} \tag{1}$$

which has only eight nonzero elements in case of macroscopically isotropic and symmetric media. The phase function for a single non-spherical particle depends on its orientation with respect to the direction of incident radiation and on the particle microphysical characteristics. In addition, the phase function varies not only with scattering angle θ but also with azimuth angle. Both these angles characterize the direction of propagation of the scattered radiation for a given particle orientation. To obtain $S_{11}(\theta)$ for Gaussian particle, the numerical integration over the whole solid angle is necessary, which is a time- and memory- consuming procedure if based on DDA.

The isolines of $S_{11}(\theta)$ drawn in Figure. 7 document differences between Gaussian particles and homogeneous spheres. In particular, the same intensities of scattered radiation are observed at different scattering angles. The isolines for homogeneous sphere are a bit shifted to the smaller scattering angles. It indicates that scattering phase function for the small Gaussian particle is slightly more prolonged in the forward direction. Translated to the scattering efficiencies, the Q_{sca} for homogeneous sphere has to be smaller than that for Gaussian particle. This perfectly coincides with results presented in Figure. 4. The opposite situation with $S_{11}(\theta)$ is found for larger particles (Figures. 8 and 9). The lobe of the scattering phase function is more elongated in case of homogeneous spheres.

These findings are important in evaluation of the bulk optical properties of a particle population. In principle, the optical response of a particle polydisperse system is given as a weighted contribution of all single particles assuming their size distribution functions.

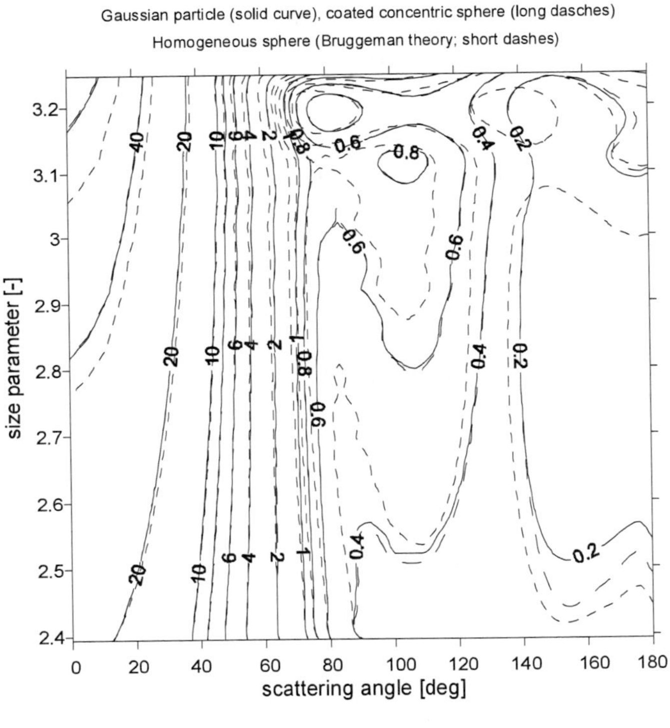

Figure. 8. The same as Figure. 7 but for particle radii $r_0 = 0.2$ μm.

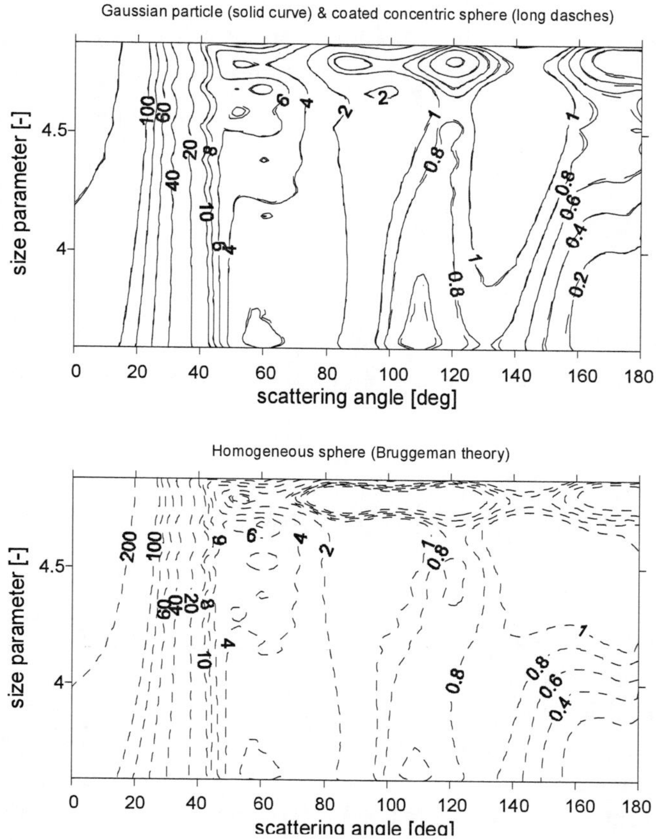

Figure. 9. The same as Figure. 7 but for particle radii $r_0 = 0.3$ μm.

4. Modeling the Bulk Optical Properties of Aerosol Ensembles

Aerosol particles scatter and absorb radiation across the whole visible and IR spectrum. Airborne particles are of diverse origin, and they have different material and chemical compositions (and optical properties), size distributions, and spatial distributions. These characteristics usually change in time. The short-term fluctuations may be due to atmospheric turbulence, while the long-term changes are related to an exchange of air masses.

On the one hand, all these aerosol characteristics (particularly size distribution and refractive index) are crucial factors affecting the transmission of optical signals in the atmosphere. It is known, that any small fluctuations of these characteristics may result in important changes of the radiation field. This is one of the reasons for systematic research of aerosols microphysical characteristics.

On the other hand, the physical processes in the atmosphere largely initiate the spatial and temporal changes of the aerosol system and these in turn influence the scattering and absorption efficiencies of the atmospheric environment.

A rapid increase of spatial concentration of aerosol particles affects the turbidity of the atmospheric environment. In a strongly polluted atmosphere the single-scattering approach becomes inapplicable for solving the light scattering problems. The total concentration of the particles N [m^{-3}] is an integral of so-called number size distribution function $f(r)$

$$N = \int_{0}^{\infty} f(r)dr \quad , \tag{2}$$

where r is the particle radius. The conventional measuring devices are unable to identify very small particles (let say the particles with radii below 0.01 μm). Therefore the cumulative concentration

$$N(r_{\min}) = \int_{r_{\min}}^{\infty} f(r)dr \tag{3}$$

is available rather than total concentration N. The concentration $N(r_{\min} > 0.01)$ show large variability mainly in the boundary layer of the atmosphere.

The vertical stratification of aerosol particles is an important factor for modeling the radiative transfer and optical properties of Earth's atmosphere. An acceptable model for vertical stratification of aerosol concentration must be satisfactory simple, but close to reality. It is commonly accepted that the concentration of optically active particles (with the size comparable to the wavelength of the incident radiation) exponentially decreases with the altitude h, so

$$N(r_{\min}, h) = N(r_{\min}, h_0)\exp\left(\frac{h_0 - h}{h_A}\right) \quad , \tag{4}$$

where h_A is the so-called scale height of aerosol atmosphere, i.e. it is an altitude up to which a homogeneous aerosol atmosphere would extend. The altitude h_A varies from 1.2 up to 1.7 km.

The most frequently employed formula in modeling the aerosol content is so-called Junge's function

$$f(r) = \frac{dN}{dr} = c\,r^{-n} \quad , \tag{5}$$

where c is a free parameter. Although this function is often recommended to fit the real size distributions (e.g. it characterizes the number size distribution of large particles very well), the amount of small particles is evidently overestimated. The fact that Junge's distribution describes only the particle system in a limited interval of particle radii

$\langle r_{\min}, r_{\max} \rangle$ must be taken into account. The applicability of the Junge's function to the particles with sizes less than 1 μm and greater than 0.1 μm was proven by Davies (1974). The further analyses of atmospheric aerosols in mountain regions have shown that a validity of the power law may cover a wide region from approximately 0.03 μm to 3 μm. However, Junge's function does not work well in the industrial agglomerations, for which the rapid exchange of the aerosol systems is typical. Under such conditions, the number size distribution either shows a complex behavior or it has an obvious mode located somewhere in the submicron range. There exist some modifications of Junge's function to cover such situations. For instance, the piecewise Junge's distribution may have a form (Zuev and Krekov, 1986)

$$f(r) = c\,r^{-n}, \quad n = \begin{cases} 0 & 0.00 < r \le 0.05\,\mu m \\ 2.1 & 0.05 < r \le 0.09\,\mu m \\ 3.4 & 0.09 < r \le 0.90\,\mu m \\ 5.3 & 0.90 \le r \le 1.00\,\mu m \\ 5.3 - 100 & 1.00 < r < \infty \end{cases} , \tag{6}$$

but these formulae are evidently mathematically impractical and thus employed in numerical modeling only sporadically. On the other hand, gamma function (Fouquard et al. 1987)

$$f(r) = A\,r^{\alpha} \exp\left(-b\,r^{\gamma}\right) , \tag{7}$$

is almost routinely used to fit the hazy conditions and can be successfully applied also to smallest fraction of the aerosol population. Typical values of the free parameters (except scaling factor A) are as follows: $\alpha = 1.0$, $b = 8.9$ [μm$^{-1/2}$], $\gamma = 0.5$ for maritime aerosols, $\alpha = 2.0$, $b = 15.1$ [μm$^{-1/2}$], $\gamma = 0.5$ for continental aerosols, and $\alpha = 2.0$, $b = 20.0$ [μm^{-1}], $\gamma = 1.0$ for stratospheric aerosols, respectively (Figure. 10).

The scaling parameter A for normalized size distribution is easily obtained in form

$$A = \frac{\gamma\, b^{(\alpha+1)/\gamma}}{\Gamma\left(\dfrac{\alpha+1}{\gamma}\right)} , \tag{8}$$

where $\Gamma(x)$ is Euler's gamma function. One of the best known size distributions is normal distribution

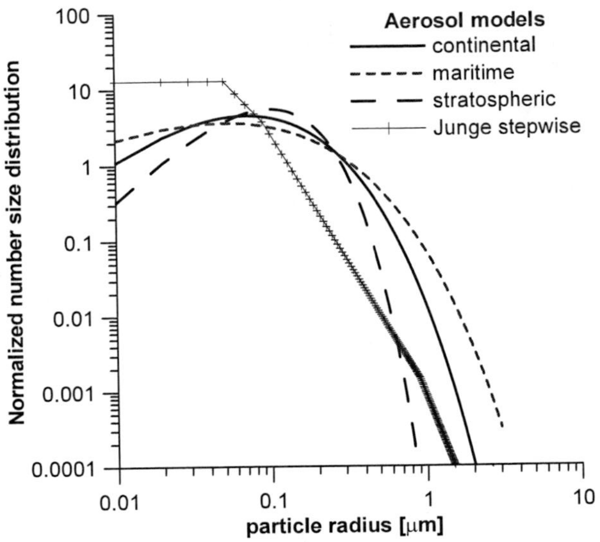

Figure. 10. Normalized number size distributions calculated by means of Eqs. (6) and (7).

$$f(r) = \frac{1}{w\sqrt{2\pi}} \exp\left\{-\frac{1}{2}\left(\frac{r-\bar{r}}{w}\right)^2\right\} \quad . \tag{9}$$

Here $\bar{r}$ is the mean radius, which for discrete size distribution is calculated from

$$\bar{r} = \frac{\int_0^\infty r\, f(r)\, dr}{\int_0^\infty f(r)\, dr} \approx \frac{1}{n}\sum_{i=1}^{n} r_i f_i \quad , \tag{10}$$

where n is a number of discrete radii and f_i is corresponding number of particles having the radius r_i. Standard deviation - a very convenient statistical property for indicating the data spread - is defined by the equation

$$w = \sqrt{\frac{1}{n}\sum_{i=1}^{n}(r_i - \bar{r})^2 f_i} \quad . \tag{11}$$

However, the mean effective radius (Volten et al, 2001)

$$\bar{r} = \frac{\int_0^\infty r^3 f(r)\, dr}{\int_0^\infty r^2 f(r)\, dr} \approx \frac{\sum_{i=1}^{n} r_i^3 f_i}{\sum_{i=1}^{n} r_i^2 f_i} \tag{12}$$

is usually applied when solving the light scattering problems for polydisperse systems of aerosol particles. The optical response of a particle on an incident electromagnetic radiation varies with effective cross section of the particle. It is not surprising when thinking in simple geometrical relations. Let us to embed a spherical particle into a non-absorbing medium and illuminate this environment by a plane wave. Imagine that the particle can be treated as an opaque target (circular disk in case of spherical shape) and thus a net rate at which electromagnetic energy crosses the particle is proportional to its cross section. Thus, in very rough approximation, the amount of energy $I_0 \pi r^2$ is removed from a beam with irradiance I_0. Of course, the reality is not so simple, and the attenuation depends on the chemical composition of the particle, its size, shape, orientation and wavelength of the incident beam. Nevertheless, it is convenient to relate the optical characteristics of the particles with their efficient cross section. Therefore the electromagnetic scattering community frequently deals with the surface distribution function (Kabanov et al., 1988)

$$s(r) = \frac{\pi r^2 . f(r)}{N} \quad , \tag{13}$$

where r is a radius of the spherical particle (in case of irregularly shaped grains r is a radius of volume equivalent sphere). The total cross section of the polydisperse system of aerosols is defined as

$$S = \int_0^\infty s(r)dr \tag{14}$$

and the total volume of the particles is

$$V = \frac{4}{3}\int_0^\infty r\, s(r)dr = \int_0^\infty v(r)dr \tag{15}$$

Here $v(r) = 4 \cdot r \cdot s(r)/3 = (4/3)\pi r^3 f(r)/N$ is so-called volume distribution function – it is the volume of particles with radii between r and $r + \Delta r$.

Distributions like (9) are symmetrical about the arithmetic mean. Normal distributions are completely determined by specifying the mean size and the standard deviation. Many size distributions are highly skewed, and can be represented by substituting some function of the radius (like in Eq. 9) or logarithm of the radius

$$f(r) = \frac{1}{\log w_g \sqrt{2\pi}} \exp\left\{-\frac{1}{2}\left(\frac{\log r - \log \bar{r}_g}{\log w_g}\right)^2\right\} \tag{16}$$

Here $\bar{r}_g = \sqrt[n]{r_1^{f_1} r_2^{f_2} r_3^{f_3} ... r_n^{f_n}}$ is the geometric mean and w_g is the geometric standard deviation (Cadle 1966). Although the log-normal distribution could be more realistic than previous functions, the Junge's or modified gamma functions represent very convenient distributions for inversion procedures (Box and Deepak 1979; Lacis and Mishchenko 1994).

In principle, many functions exist which can successfully simulate real aerosol size distributions. Nevertheless, only some of them can be written in mathematically easy formulae and only limited number of these functions can be used in analytical solutions of light scattering problems. This is a reason for introducing just the functions listed above.

The form of a size distribution may quickly change with varying relative humidity R_H because solid particles rapidly grow in a humid environment. According to Zuev and Krekov (1986) the growth rate of the particles is controlled by a quotient $\overline{\chi}$

$$r = r_0 \sqrt[3]{1 + \frac{\rho_0}{\rho_w} \overline{\chi} \frac{R_H}{1 - R_H}} \quad , \tag{17}$$

where $\overline{\chi}$ depends on the type of aerosol, r_0 and ρ_0 are initial particle radius and density, respectively, and ρ_w is the density of the water. For small sub-micrometer-sized particles, a notable size increase (in about 25%) can be observed at $R_H > 75\%$ (Hänel 1976). On the other hand, the size distribution is mostly unaffected by relative humidity when R_H is smaller than 70%. Nevertheless, the effective refractive index of aerosol particles may undergo certain changes already at $R_H \geq\sim 60\%$. Effective medium theory (Chýlek and Videen 1998) suggests that under elevated relative humidity conditions the original refractive index 1.55 of aerosol particles can be reduced to a value of 1.50. Then an estimated influence of R_H on aerosol extinction coefficient can be about 10-15% for non-absorbing particles (Kocifaj and Horvath 2004). The high R_H may apparently influence the particle morphology - an irregularly shaped solid particle could become spherical or at least near spherical in the process of humidification.

The bulk optical response of a system of non-interacting particles can be obtained by integrating the optical responses of single particles over all sizes where $f(r)$ is assumed to be a weighting function. For instance, the aerosol optical thickness τ_A is a product of the cross section for extinction ($\pi r^2 Q_{ext}$) and the size distribution function $f(r)$, i.e.

$$\tau_A(\lambda) = \pi \int_0^\infty Q_{ext}(r, \lambda) r^2 f(r) dr \tag{18}$$

Applying the anomalous diffraction approximation (van de Hulst, 1981) the efficiency factor for extinction can be obtained in the analytical form

$$Q_{ext}(y) = \frac{2m-1}{m} \left[2 - \frac{4}{y} \sin y + \frac{4}{y^2}(1 - \cos y) \right] \tag{19}$$

where $y = 2x(m-1)$ and $x = 2\pi r / \lambda$ is so-called size parameter. Assuming the size distribution of non-absorbing aerosol particles is given by Eq. (7) and $\alpha = 2$; $\gamma = 1$, the optical thickness τ_A can be also found in an analytical form

$$\tau_A(Y) = \frac{2A\pi}{Y^2}\left\{\frac{1}{b^3} + \frac{240bY^4 - 40b^3Y^2 - b^5}{(b^2+4Y^2)^4} + \frac{24Y^2}{b^5}\right\} \quad ,(20)$$

where $Y = y/(2r)$. Using the phase function S_{11} for single particles the normalized bulk phase function of aerosol system with surface size distribution (Eq. 13) can be determined as follows

$$p(\theta) = \left(\frac{\lambda^2}{2\pi}\right)\int_0^\infty \frac{\langle S_{11}\rangle(x,\theta)}{\pi r^2} s(r)\,dr \quad ,(21)$$

where $\langle S_{11}\rangle$ is the S_{11} averaged over all particle orientations, and θ is the scattering angle.

5. Diffuse and Direct Solar Radiation in Plan-Parallel Turbid Atmosphere

The modeling of the radiation field in the Earth's atmosphere includes multiple scattering and requires detailed information on the physical and optical properties of aerosol particles. Knowledge on the optical properties of diffuse environment is important not only for determination of radiative balance of different atmospheric layers, but also for correct interpretation of remote sensing measurements.

Light scattering in an inhomogeneous medium is a process which is not easy to formulate mathematically. Modeling the visible radiation penetrating the Earth's atmosphere with optical thickness greater than 0.3 is quite complex because of multiple scattering (Wauben, 1992).

Several approaches can be applied to solve the radiative transfer in an environment with multiple scattering. The method of successive scattering orders has a simple concept (Wendisch and von Hoyningen-Huene, 1991) while it is easy to implement and provides a self-explanatory outputs. However, it is slowly convergent for optically dense environments. Traditionally the "doubling or adding" method is applied because of rapid convergence, but this is not suitable for any geometry of the scattering problem.

In general, the spectral radiance I is to be determined when solving the radiative transfer in plan-parallel atmosphere (Greenwald a Stephens, 1988)

$$\mu \frac{dI(\tau;\mu,\varphi)}{d\tau} = -I(\tau;\mu,\varphi) + J(\tau;\mu,\varphi) \quad (22)$$

where $\mu = \cos\theta$ is the cosine of the zenith angle, φ is the azimuth angle, τ is altitude-dependent optical thickness of the atmosphere, and J is so-called source function. The source function comprises all internal sources of radiation, hence the scattered and reflected radiances too. In principle, J and I are closely interrelated - part of the source function in an integral product of the scattering phase function $p(\tau;\mu,\varphi)$ and $I(\tau;\mu,\varphi)$. The Eq. (22) is then an integro-differential equation which can be solved only numerically. Once the source function is known, the radiative fluxes at any altitude can be calculated (Herman et al., 1995, Liang and Strahler, 1995). The total radiative flux $F(\tau)$ is a cosine-weighted integral of spectral radiance

$$F(\tau) = \int_0^{2\pi}\int_{-1}^{1} I(\tau;\mu,\varphi)\,\mu\, d\mu\, d\varphi \tag{23}$$

where the integration runs over the whole solid angle. Note, that the downward irradiance on a horizontal surface is obtained when integrating over $\mu \in \langle -1,0 \rangle$. Logically, the upward irradiance can be calculated if the integration range is limited to $\mu \in \langle 0,1 \rangle$.

An easiest way for determining the diffuse and direct solar radiative fluxes is to use the two-stream approximation (Ceballos, 1988, Xiang et al 1994). For instance, radiative fluxes in the plan-parallel atmosphere can be computed by Eddington's approximation with accuracy obviously depending on the chosen grid, i.e. on the number of model atmospheric layers. In comparison with more accurate four-stream methods the maximum error of the two-stream approximation is about 10% for radiance data. Such an error margin is acceptable to evaluate an effect of relative humidity on radiative fluxes close to the Earth's surface. We have used a modified implementation of two-stream Eddington's approximation. Compared with conventional algorithms (LOWTRAN 7, Lenoble, 1995), the modified implementation allows user-specific configuration, thus being well scalable.

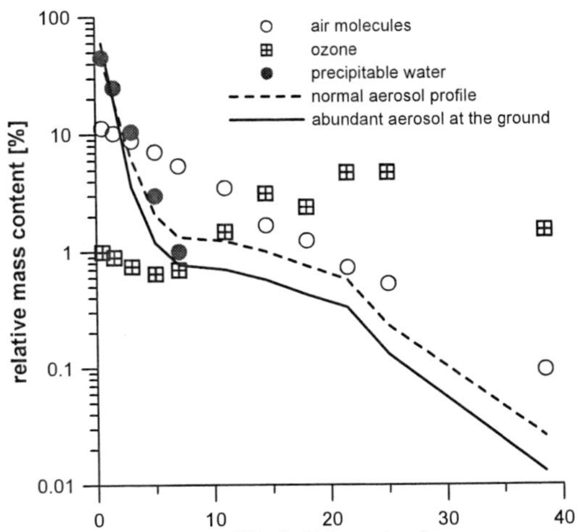

Figure. 11. Relative mass contents of some atmospheric constituents.

We have assumed the lambertian spectrally-dependent surface albedo a_λ and altitude-dependent aerosol size distribution function (preconfigured models comprise Junge's function, Gamma or Gauss functions, and data entry). Note, that aerosol particles are concentrated mainly in the lower troposphere, and their mass rapidly drops as the altitude increases (Figure. 11).

The Rayleigh scattering by air molecules, and the attenuation of solar radiation due to aerosol particles, ozone, and precipitable water are taken into consideration when simulating radiative fluxes under elevated humidity conditions. In many practical applications the flux density of photosynthetically active radiation (PhAR) need to be evaluated. It comprises not only the downward diffuse $F_D^{\downarrow}$ and direct solar $F_S^{\downarrow}$ irradiances, but also an upward diffuse irradiance $F_D^{\uparrow}$. The difference of downward and upward radiative fluxes is usually interpreted in terms of radiative balance of individual atmospheric layers. At the ground it relates to the energetic behavior of the system Earth-atmosphere.

Numerical modeling of radiative transfer is reduced to one-dimensional problem. We assume the plan-parallel atmosphere with lambertian surface (i.e. a perfectly diffusing surface that reflects incident radiation in accordance with the Lambertian cosine law). Although the surface albedo a_λ usually depends on the wavelength of an incident radiation, the averaged equivalent $\bar{a}$

$$\bar{a}(\mu) = \frac{\int_{400\,nm}^{700\,nm} a_\lambda(\mu)\left[F_D^{\downarrow}(\mu) + F_S^{\downarrow}(\mu)\right]d\lambda}{\int_{400\,nm}^{700\,nm} \left[F_D^{\downarrow}(\mu) + F_S^{\downarrow}(\mu)\right]d\lambda} \qquad (24)$$

is traditionally used in numerical simulations. Here the integration runs over PhAR. Spectral albedo for some types of surfaces may differ significantly (Figure. 12). Kozhaev and Shifrin (1972) showed that a_λ for a fresh snow is 0.8 at λ = 400 nm but only 0.65 at λ = 700 nm. The albedo of a dirty snow is almost independent of wavelength (it is roughly 0.7 throughout the PhAR). In general, the spectral albedo may depend on the cosine of solar zenith angle μ (as indicated in Eq. 24). However, Kastrov (1979) showed that the values of $\bar{a}(\mu)$changes only within a few percent in the PhAR. We have verified the applicability of approximate equation (24) for number of situations and for all three surfaces presented in Figure. 12. In particular, the aerosol microphysical characteristics have been varied to simulate strongly absorbing as well as slightly-absorbing material compositions and size distributions after Eqs. 6. and 7. The PhAR irradiances $F_D^{\uparrow}$ become overestimated by 10% for soil surfaces and about 30% for vegetation coverage if spectrally-dependent albedo a_λ is replaced by $\bar{a}$. The water causes underestimation of $F_D^{\uparrow}$ by a few percent.

The irradiances $F_D^{\downarrow}$, $F_S^{\downarrow}$, and $F_D^{\uparrow}$ are calculated under different relative humidity conditions. The aerosol characteristics (size distribution, refractive index, total aerosol optical depth) evolving in the process of humidification are modelled in accordance to Shettle and Fenn (1979). The total amounts of 0.3 and 3.0 atmospheric centimetres are considered for ozone and precipitable water, respectively. The downward and upward diffuse irradiances for vegetation-covered surfaces are computed numerically and documented in Figures. 13-14. It is shown that diffuse radiation reaching the Earth's surface monotonously increases with growing R_H. The largest growth rate of $F_D^{\downarrow}$ is observed at elevated relative humidity (above 75%), while $F_D^{\downarrow}$ changes only slightly when R_H is below 50% (Figure. 13). In contrast, the upward diffuse radiation is only slightly sensitive to R_H. In the process of humidification the upward irradiance is a decreasing function of R_H, but $F_D^{\uparrow}$ changes only within 5% (Figure. 14).

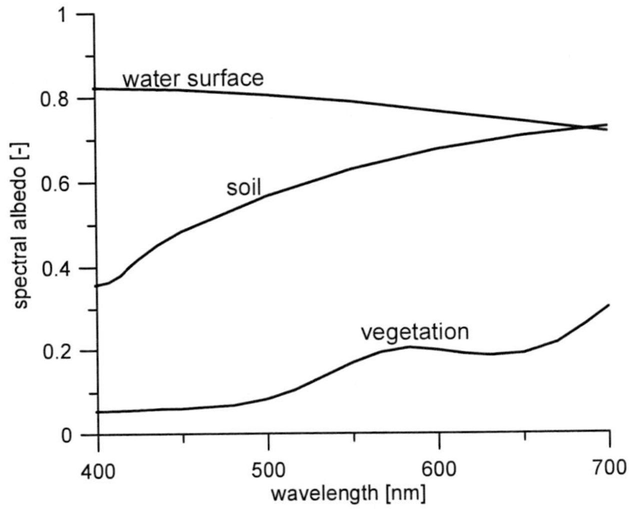

Figure. 12. Spectral behavior of Earth's albedo for some types of surfaces (after Mishev1985).

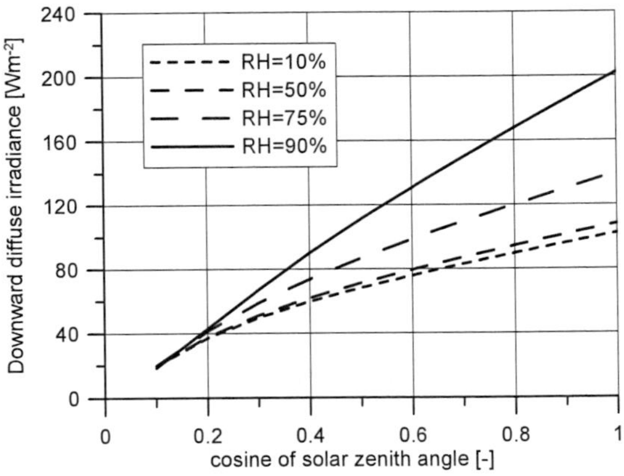

Figure. 13. Downward diffuse irradiance (400-700 nm) under evolving relative humidity (RH) conditions.

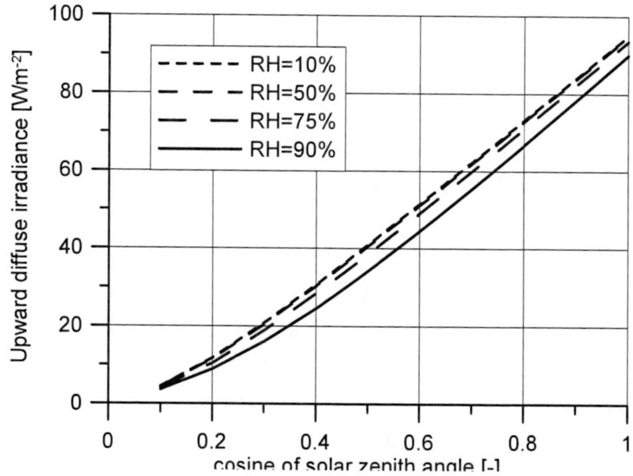

Figure. 14. Upward diffuse irradiance (400-700 nm) under evolving relative humidity (RH) conditions.

The direct solar irradiation $F_S^{\downarrow}$ drops down as R_H increases, however, it shows more steep behavior than in case of $F_D^{\uparrow}$. It can be explained by large values of aerosol optical thickness under elevated humidity. The larger R_H, the greater particle sizes and thus extinction cross sections.

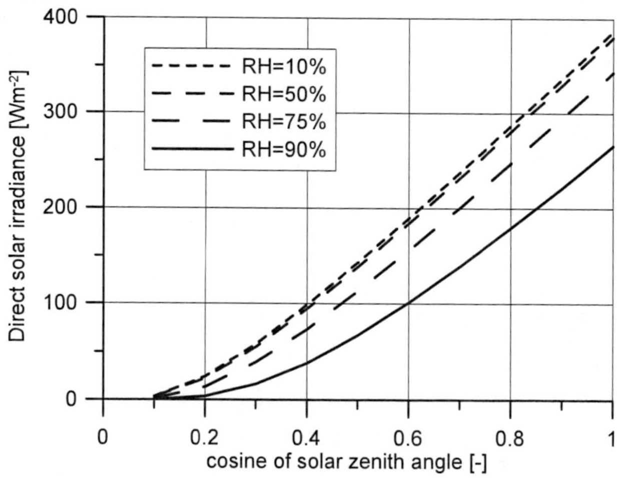

Figure. 15. Downward direct solar irradiance (400-700 nm) under evolving relative humidity (RH) conditions.

6. Concluding Remarks

Aerosol particles of different origin are constantly present in the atmosphere. On the one hand, the aerosol particles influence the physical processes in the atmosphere, and, on the other hand these processes trigger changes in aerosol system. This is a kind of inherent regulation, in which the aerosol is one of catalyst factors. There is no doubt that atmospheric aerosol attenuates the intensity of direct solar radiation across the whole visible spectrum and subsequently contributes to the global climate change through the radiation forcing (Horvath et al., 2002).

Information on aerosols is acquired by optical (contactless) methods as well as mechanically (using various filtration techniques). However, filtration can provide only information about the so-called dry aerosol fraction - hence the solid dust grains. It is well recognized that airborne aerosol particles are humid and water usually forms the shell coating of the solid nuclei. The water content can be quite large under elevated relative humidity. When the aerosol particles grow in the process of humidification, their refractive indices monotonously approach those of water drops. These changes markedly influence the optical behavior of aerosol population as well as the optical response of an ambient atmospheric environment. It can be stated that humidity influences the complex processes of aerosol formation and evolution. The knowledge on aerosol microphysical properties and their alteration in the process of humidification is therefore necessary for correct interpretation of measured optical data as well as for improved estimations of radiative impacts of aerosols on climate. In this chapter we discussed some effects of relative humidity on the optics of single particles. In addition, the radiative fluxes at the ground were simulated for somemodel situations.

ACKNOWLEDGMENT

This work has been supported by the Slovak Research and Development Agency under the contract No: APVV SK-CZ-0015-09.

REFERENCES

Bohren, C. F. & Huffman, D. R. (2004). *Absorption and Scattering of Light by Small Particles*, Wiley-VCH, Weinheim.

Borde, R. & Verdebout, J. (2003). Remote sensing of aerosols optical thickness over various sites using SeaWiFS or VEGETATION and ground measurements. *Remote Sensing of Environment*, *86*, 42-51.

Box, M. A. & Deepak, A. (1979). Retrieval of aerosol size distributions by inversion of simulated aureole data in the presence of multiple scattering, *Applied Optics*, *18*, 1376-1382.

Bruggeman, D. A. G. (1935). Calculation of various physics constants in heterogeneous substances. I. Dielectricity constants and conductivity of mixed bodies from isotropic substances. *Ann Phys*., *24*, 636-664.

Bukowiecki, N., Kittelson, D. B., Watts, W. F., Burtscher, H., Weingartner, E. & Baltensperger, U. (2002). Real-time characterization of ultrafine and accumulation mode particles in ambient combustion aerosols. *Journal of Aerosol Sciences*, *33*, 1139-1154.

Cadle, R. D. (1966). Particles in the atmosphere and space. *Reinhold publishing corporation*, New York.

Charalampopoulos, T. T. (1992). Morphology and dynamics of agglomerated particulates in combustion systems using light scattering techniques. *Progress in Energy and Combustion Science*, *18*, 13-45.

Ceballos, J. C. (1988). On two-flux approximations for short-wave radiative transfer in the atmosphere. *Beiträge zur Physik der Atmosphäre*, *61*, 10-22.

Chýlek, P. & Videen, G. (1998). Scattering by a composite sphere and effective medium approximations. *Optics Communications*, *146*, 15-20.

Davies, C. N. (1974). Size distribution of atmospheric particles. *Journal of Aerosol Science*, *5*, 293-300.

Draine, B. T. (1988). The discrete-dipole approximation and its application to interstellar graphite grains. *Journal of Astrophysics*, *333*, 848-872.

Esposito, F., Leone, L., Pavese, G., Restieri, R. & Serio, C. (2004). Seasonal variation of aerosol properties in South Italy: a study on aerosol optical depths, Angstrom turbidity parameters and aerosol size distributions. *Atmospheric Environment*, *38*, 1605-1614.

Farafonov, V. G., Voshchinnikov, N. V. & Somsikov, V. V. (1996). Light scattering by a core-mantle spheroidal particle. *Applied Optics*, *35*, 5412-5425.

Fouquard, Y., Bonnel, B., Chaoui, M. & Santer, R. (1987). Observations of Saharan aerosols: results of ECLATS field experiment. Part I: Optical thickness and aerosol size distribution. *Journal of Climate and Applied Meteorology*, *26*, 28-37.

Golovko, V. F. (2000). Dispersion formula and continuous absorption of water vapor. *Journal of Quantitative Spectroscopy and Radiative Transfer*, *65*, 621-644.

Greenwald, T. J. & Stephens, G. L. (1988). Application of a doubling-adding radiation model to visibility problems. Final Report, CIRA, Colorado State University, Foothills Campus, *Fort Collins Colorado*, 80523.

Hansen, J., Sato, M. & Ruedy, R. (1997). Radiative forcing and climate response. *Journal of Geophysical Research*, *102*, 6831-6864.

Hänel, G. (1976). The properties of atmospheric aerosol particles as functions of the relative humidity at thermodynamic equilibrium with the surrounding moist air. *Advances in Geophysics*, *19*, 74-183.

Herman, B. M., Caudill, T. R., Flittner, D. E., Thome, K. J. & Ben-David, A. (1995). Comparison of the Gauss-Seidel spherical polarized radiative transfer code with other radiative transfer codes. *Applied Optics*, *34*, 4563-4572.

Horvath, H., Kasahara, M. & Pesava, P. (1996). The size distribution and composition of the atmospheric aerosol at a rural and nearby urban location. *Journal of Aerosol Science*, *27*, 417-435.

Horvath, H., Arboledas, L. A., Olmo, F. J., Jovanović, O. & Gangl, M. (2002). Optical characteristics of aerosol in Spain and Austria and its effect on radiative forcing. *Journal of Geophysical Research*, *107*, 4386-4404.

van de Hulst, H. C. (1981). *Light scattering by small particles*. Wiley, New York.

Kabanov, M. V., Panchenko, M. V., Pkhalagov, Yu. A., Veretennikov, V. V., Uzhegov, V. N. & Fadeev, V. Ya. (1988). *Optical properties of the maritime smokes*. Nauka, Novosibirsk, (in Russian)

Kahnert, F. M. (2003). Numerical methods in electromagnetic scattering theory. *Journal of Quantitative Spectroscopy and Radiation Transfer*, 79-80, 775-825.

Kastrov, V. G. (1979). *Selected papers of the physics of atmosphere. Gidrometeoizdat*, Leningrad.

Kato, S., Ackerman, T. P., Clothiaux, E. E. & Mathe,r, J. H. (1997). In *Current Problems in Atmospheric Radiation*: Proc. Int. Radiation Symp. *Fairbanks*, Alaska, 19-24 August 1996. A. Deepak, Hampton, VA.

Kent, G. S., Yue, G. K., Farrukh, U. O. & Deepak, A. (1983). Modeling atmospheric aerosol backscatter at CO2 laser wavelengths. 1: aerosol properties, modeling techniques, and associated problems. *Applied Optics*, *22*, 1655-1665.

Kocifaj, M. & Horvath, H. (2004). Attenuation of optical signals in the atmosphere: An analytical approach. In: J. Pinker, (Ed.), Applied Electronics 2004. *University of West Bohemia*, Pilsen, 109-112.

Kocifaj, M., Kundracik, F. & Videen, G. (2008). Optical properties of single mixed-phase aerosol particles. *Journal of Quantitative Spectroscopy and Radiative Transfer*, *109*, 2108-2123.

Krueger, A. J. & Mizner, R. A. (1976). A mid-latitude ozone model for the 1976 US standard atmosphere. *Journal of Geophysical Research*, *81*, 4477.

Lacis, A. A. & Mishchenko, M. I. (1994). Climate forcing, climate sensitivity, and climate response: A radiative modeling perspective on atmospheric aerosols. In: R. L. Charlson, & J. Heintzenberg, (Eds.), *Aerosol forcing of climate*. John Wiley & Sons.

Lahaye, J. & Prado, G. (1981). Morphology and internal structure of soot and carbon blacks. In Particulate Carbon (Edited by D. C. Siegla, & G. W. Smith,), 31-55. *Plenum*, New York.

Latha, K. M. & Badarinath, K. V. S. (2005). Spectral solar attenuation due to aerosol loading over an urban area in India. *Atmospheric Research*, *75*, 257-266.

Liang, S. & Strahler, A. H. (1994). Four-stream solution of atmospheric radiative transfer over a non-Lambertian surface. *Applied Optics*, *33*, 5745-5753.

Mie, G. (1908). Beiträge zur Optik trüben Medien, speziell kolloidaler Metallösungen, *Annals of Physics*, *25*, 377-445.

Mishev, D. (1985). *Remote sensing of the Earth from the space*. Mir, Moscow.

Moya, M., Grutter, M. & Báez, A. (2004). Diurnal variability of size-differentiated inorganic aerosols and their gas-phase precursors during January and February of 2003 near downtown Mexico City. *Atmospheric Environment*, *38*, 5651-5661.

Niemi, J. V., Tervahattu, H., Vehkamäki, H., Kulmala, M., Koskentalo, T., Sillanpää, M. & Rantamäki, M. (2004). Characterization and source identification of a fine particle episode in Finland. *Atmospheric Environment*, *38*, 5003-5012.

Park, S. S., Kim, Y. J. & Fung, K. (2001). Characteristics of PM2:5 carbonaceous aerosol in the Sihwa industrial area, South Korea. *Atmospheric Environment*, *35*, 657-665.

Pickering, K. E., Thompson, A. M., Wang, Y. S., Tao, W. K., McNamara, D. P., Kirchhoff, V., Heikes, B. G., Sachse, G. W., Bradshaw, J. D., Gregory, G. L. & Blake, D. R. (1996). Convective transport of biomass burning emissions over Brazil during TRACE A. *Journal of Geophysical Research-Atmospheres*, *101*, 23993.

Shettle, E. P. & Fenn, R. W. (1979). *Models for the Aerosols of the Lower Atmosphere and the Effects of Humidity Variations on Their Optical Properties*. AFGL-TR-79-0214, 20 Sept 1979, ADA085951.

Shifrin, K. S. & Kozhaev, D. A. (1972). Determination of the albedo of extensive territories bz angular structure of cloud sky brightness, In: *Optics of the ocean and atmosphere*, Ed. K. S. Shifrin, Nauka, Leningrad.

Sokolik, I. N. & Toon, O. B. (1999). Incorporation of mineralogical composition into models of the radiative properties of mineral aerosol from UV to IR wavelengths. *Journal of Geophysical Research, 104*, 9423-9444.

Volten, H., Muñoz, O., Hovenier, J. W., de Haan, J. F., Vassen, W., van der Zande, W. J. & Waters, L. B. F. M. (2005). WWW scattering matrix database for small mineral particles at 441.6 and 632.8 nm. *Journal of Quantitative Spectroscopy and Radiative Transfer, 90*, 191-206.

Volten, H., Muñoz, O., Rol, E., de Haan, J. F., Vassen, W., Hovenier, J. W., Muinonen, K. & Nousiainen, T. (2001). Scattering matrices of mineral particles at 441.6 nm and 632.8 nm. *Journal of Geophysical Research, 106*, 17375-17401.

Wauben, W. M. F. (1992). Multiple scattering of polarized radiation in planetary atmospheres. *Vrije Universiteit*, Enschede.

Wendisch, M. & von Hoyningen-Huene, W. (1991). High speed version of the method of 'successive order of scattering' and its application to remote sensing. *Beiträge zur Physik der Atmosphäre, 64*, 83-91.

Wickramasinghe, N. C. (1963). On graphite particles as interstellar grains, II. *Monthly Notices of the Royal Astronomical Society, 126*, 99-114.

Yang, P., Liou, K. N., Mishchenko, M. I. & Gao, B. C. (2000). Efficient finite-difference time-domain scheme for light scattering by dielectric particles: application to aerosols. *Applied Optics, 39*, 3727-3737.

Xiang, X., Smith, E. A. & Justus, C. G. (1994). A rapid radiative transfer model for reflection of solar radiation. *Journal of the Atmospheric Sciences, 51*, 1978-1988.

Zuev, V. E. & Krekov, G. M. (1986). *Optical models of the atmosphere*. Leningrad: Gidrometeoizdat (in Russian).

In: Relative Humidity: Sensors, Management...
Editor: John M. Castillo

ISBN: 978-1-61761-734-8

Chapter 5

EFFECTS OF HEAT STRESS ON REPRODUCTIVE EFFICIENCY IN HIGH PRODUCING DAIRY COWS

***P. Santolaria*[1], *F. López-Gatius*[2], *I. García-Ispierto*[2] *and J. Yániz*[1]**

[a] Institute of Environmental Science and Animal Production Department. Escuela Politécnica Superior. University of Zaragoza, Ctra. Cuarte s/n. 22071. Huesca. Spain
[b] Department of Animal Production, University of Lleida, Lleida, Spain

ABSTRACT

Climate plays an important role in modulating reproductive success in cattle. Lactating dairy cows are under high metabolic demands due to high milk production and in this situation stressful factors such as climate conditions can compromise reproductive functions. Environmental temperature, radiant energy, relative humidity, wind speed and rainfall could contribute to the degree of ambient stress. To assess the influence of relative humidity (RH) and ambient temperature, the use of a temperature- humidity index (THI) which incorporates the effects of both is widespread. Several studies have related the THI index with fertility. In these studies, it is noteworthy that fertility decreased from 34% to 25% when THI value were higher than 75 on the day of insemination, or that THI values higher than 80 three days before insemination decrease fertility from 30-34% to 23%. On the other hand, prenatal loss is also an important factor affecting reproductive performance in high producing dairy cattle and high values of temperature-humidity index (THI) on days 21 to 30 of gestation have been found to be a risk factor for subsequent early foetal loss. Also, cumulative stressful climatic conditions and episodes of acute variation of climatic conditions in the first 40 days of gestation can compromise the success of gestation during both the cool and warm period of the year. A one-unit increase in the cumulative number of hours with THI > 85 during the days 11-20 of gestation period caused a 1.57-fold increase in the pregnancy loss, whereas the likelihood of foetal loss increased by a factor of 1.16 for each additional episode of acute increases in maximum THI values during gestation days 0-40. Finally, it is reported that during the second trimester of gestation, increased days with low RH in N. caninum-infected parous cows and heifers can compromise the success of gestation in a dry environment.

1. Introduction

The impact of environmental conditions on animals has been alluded to since antiquity. In On Airs, Waters and Places, Hippocrates in the 5th century B.C. saw that cattle raised in the Near East were more prolific than European cattle because of the temperate climate. The last century high temperature had been related to decrease fertility for the first time (Erb et al., 1940; Seath and Staples, 1941) and since then, many researchers have focused on climate conditions and animal production. Nowadays the threat of climatic change has exacerbated the problem. Thus, a great effort has been made to foresee the possible negative effects in animal breeding, and especially in dairy cattle. However, little attention has been paid to the thermoregulatory ability of the high producing cow to deal with environmental conditions.

Temperature gradients have special relevance in the mammalian reproductive system. Domestic ruminants maintain homoeothermic or deep body temperature at about 38.6°C, within a wide thermoneutral range where nutrition and genetic interactions can proceed regardless of air temperature. The most comfortable temperatures for dairy cows ranged from 5 to 25°C, which is the thermal comfort zone (Mc Dowell, 1972). In this zone, no additional energy above maintenance is expended to heat or cool the body, so minimal physiological cost and maximum productivity are achieved (Folk, 1974). In countries with tropical and subtropical climates, the 25°C temperature is exceeded in many days in summer, and even in spring or autumn. Therefore, the problem of heat stress in not confined in tropical regions of the world and has a great impact on farm economy. Moreover, European cattle breeds (Bos taurus) are less tolerant to heat stress than zebu (Bos indicus). Such stress can disrupt both the physiology and productive performance of dairy cows. The most important effect of heat stressed cows is the increase in infertility (DeRensis and Scaramuzzi, 2003; López-Gatius, 2003).

Generally, the thermal comfort zone ranges from lower critical temperature (LCT) to upper critical temperature (UCT). This range depends on age, breed, food intake, diet composition, previous temperature acclimation or acclimatisation status, production, specific housing and pen conditions, tissue insulation (fat, skin), external insulation (coat), and behaviour of an animal (Yousef, 1985). Cattle body temperatures can vary by several degrees during the day. The coolest body temperatures occur in the early morning while the warmest occur around 6 p.m. (Beede et al., 1991). When environmental factors such as temperature and humidity rise, the UCT physiological mechanisms start to control internal body temperatures. Dairy cows dissipate heat through conduction, convection, radiation and evaporative cooling. Conduction is based on the principal that heat flows from warm to cold. This method of heat loss requires that cows have physical contact with surrounding objects. Cooling by convection occurs when air next to the skin is replaced with cooler air. Radiation of body heat occurs when the ambient temperature is significantly cooler than cows. At cool temperatures, dairy cattle are efficient at radiating heat. Finally, evaporative cooling occurs when sweat or moisture is evaporated from the skin or respiratory tract. This explains why dairy cattle sweat and increase respiratory rate during heat stress. High humidity limits the ability of the cow to take advantage of evaporative cooling (Beede et al., 1991; Collier et al., 2006). At LCT, normal metabolic processes supply enough heat to maintain body temperature. Within the thermal comfort zone, animals may modify their behaviour. For example, cows protect against wind and create a thick hair coat for winter. However, below the lower limit of the thermoneutral zone, in the LCT, the animal experiences cold stress. To overcome cold stress, the animal must increase its metabolic rate to supply more body heat. This increases dietary requirements, particularly for energy. Factors that modify effective temperature during winter are cold, wind, snow, rain and fog. Cold wind passing over an animal draws heat away from it much more quickly than warmer air. This effect is greater if the animal is wet (McDowell, 1972; Broucek et al., 1991).

High milk producing dairy cows are always under the metabolic stress of the production. Therefore, higher producers are more sensitive to other external stress than lower performers, as milk production involves metabolic heat. So, although there are still few countries where cold stress can affect reproduction; heat stress emerges as one of the most important stressors of dairy cattle worldwide.

Climate plays an important role in reproductive and productive function, especially heat stress on high performance dairy cows. The most important effect of heat stress is reduction in fertility and increased pregnancy loss, factors already altered in modern dairy cattle. This chapter will focus on those topics.

2. Measuring Heat Stress

Effective environmental temperature (EAT) can be affected by 4 factors (Buffington et al, 1981):

- Air temperature
- Solar radiation
- Air movement
- Relative humidity

However, attempts to quantify EAT have had limited success. Combining these four parameters in an index has not been possible to date. A very recent index emerged to combine all four variables (Gaughan et al., 2008), but more studies should be performed to assess the effectiveness of this new index in dairy cattle. The temperature–humidity index (THI) incorporates the effects of both ambient temperature and relative humidity (RH) in an index (Thom, 1958). This index was created by Thom (1958) to analyse the comfort thermal sensation in humans, and the Livestock Conservation Institute later observed that this index was a good heat stress predictor also for dairy cows.

The mean and maximum THI were fitted to the following equations (Thom, 1959; McDowell et al., 1972):

$$\text{Mean THI}= (0{,}8\text{xmean T}+ (\text{mean RH } (\%)/100) \text{ x } (\text{mean T}-14{,}4)+46{,}4)$$

$$\text{Maximum THI}= (0{,}8\text{xmaximum T}+ (\text{minimum RH } (\%)/100) \text{ x } (\text{maximum T}-14{,}4)+46{,}4)$$

When T is temperature and RH is relative humidity.

Maximum THI is closer to the effect of climate on the biological systems than the remaining parameters. High temperatures are always accompanied by low relative humidity (normally in the midday) while hours with lower temperatures are always accompanied by high relative humidity concentration (dew). Thus, maximum THI has recently become one of the most used indexes to assess the impact of heat stress in dairy cows worldwide (Hahn, 1969; Fuquay 1981).

Wiersama, 1990, established four temperature-humidity ranges of concern (Table 1).

- Under 72 units: no heat stress
- 72-78 units: mild heat stress
- 78-82: moderate heat stress
- Above 82: severe heat stress

This classification has been widely used in animal production, but some clarification is needed. Recent research shows that increasing milk production from 35 to 45litres/day reduces the threshold temperature for heat stress by 5°C (Berman, 2005). Consequently, a THI of 72 may under-estimate heat load in high-yielding Holstein-Friesian cows. Also, THI does not account for solar radiation or air movement. Both factors, along with air temperature and relative humidity, determine the heat gained and lost by the cow. Finally, THI does not take into account accumulation of heat load over time. Despite these limitations, THI is still the most used index worldwide due to its simplicity and proven value for dairy cattle.

Table 1. Temperature Humidity Index (THI), calculated from Temperature (°C) and Relative Humidity (%), highlighting its effect on cow stress and hence performance. Adapted from F. Wiersama, 1990. Dep. Agric. Eng., University of Arizona, Tucson, Arizona, USA

	0	5	10	15	20	25	30	35	40	45	50	55	60	65	70	75	80	85	90	95	100
22																				72	72
23																72	72	73	73	74	74
26									72	73	73	74	74	75	76	76	77	77	78	78	79
29				72	73	73	74	75	75	76	77	78	78	79	80	80	81	82	83	83	84
32	72	73	74	75	76	77	78	79	79	80	81	82	83	84	85	86	86	87	88	89	90
35	75	76	77	78	79	80	81	82	83	84	85	86	87	88	89	90	91	92	93	94	95
38	77	79	80	81	82	83	86	86	87	88	89	90	92	93	96	95	96	97	99		
41	80	81	82	85	86	86	88	89	90	91	93	94	95	97	98	99					
43	81	83	86	86	87	88	90	91	92	94	96	97									
44	82	83	85	86	88	89	91	93	94	95	96	98									

72-78 Mild stress; 78-89 Severe stress; 89-98 Very severe stress

2. Effect of Climate Factors on Fertility and Ovarian Disorders

In recent decades, declining fertility and increasing milk production in dairy cattle have been reported worldwide. Causes of fertility failure are of multifactorial origin, and the relevance of different factors varies depending on the geographical area of study and the herd. The roles of genetics, environment, nutrition and management practices have been described in several reports (for reviews see Lucy 2001; Dobson et al. 2007; Lucy 2007). Moreover, in warm countries such as Spain, summer heat stress is a major factor affecting fertility in high-producing dairy herds (Labèrnia et al. 1998; López-Gatius 2003; García-Ispierto et al. 2007a).

In a recent review, de Rensis and Scaramuzzi (2003) identified two largely independent pathways by which heat stress leads to infertility. The first is a direct effect of hyperthermia on the reproductive axis, ovary and uterus physiology. For example, it is possible that, under heat stress conditions, the oviduct or the intrauterine environment may be compromised with a decrease in blood flow and an increase in their internal temperature (Gwazdauska et al., 1975). Exposure of gametes and early embryo to high temperatures after AI could compromise their survival and success of fertilisation. The second pathway is though the effect of heat stress on appetite and consequently, on dry matter intake (Hansen, 1997; Wilson et al, 1998; Drew, 1999). The consequence is a worsening energy balance and since the postpartum dairy cow tends to be in negative balance, the consequences of heat stress on fertility are more likely to be severe. The effects of heat stress may be exacerbated in high producing dairy cows compared with low producers, because metabolic demands placed upon high-yielding dairy cows are greater and should be added to the metabolic stress related to high temperatures. However, in a large-scale study in high-producing dairy herds, we could not detect any effects of milk production at AI, or interaction between milk production and season, on conception rate (López-Gatius et al. 2003).

Table2. Odds ratios of the variables included in the final logistic regression model for pregnancy rate

Factor	Class	n	% fertility	Odds ratio	95% Confidence Interval	P
Maximum THI Day 3 before AI	<70	2292/6659	34.4	1.5	1.1-1.9	0.05
	71-75	464/1407	33.0	1.5	1.1-1.9	0.05
	76-80	392/1279	30.6	1.5	1.2-1.9	0.02
	81-85	246/1071	23.0	1.1	0.9-1.1	0.46
	>86	119/548	21.7	Reference		
Maximum THI of AI day	<70	2316/6630	34.9	1.7	1.1-2.3	<0.001
	71-75	466/1400	33.3	1.5	1.2-2.0	0.002
	76-80	225/1306	25.7	1.1	0.9-1.4	0.44
	81-85	278/1047	26.6	1.3	1.0-1.7	0.04
	>86	121/581	20.8	Reference		

Likelihood ratio test = 491.07; 70df, P=0.0001

Hosmer and Lemeshow Goodness-of-fit test = 2.12; 8 df, P = 0.98 (the model fits).

In an exhaustive study on 10965 inseminations, García Ispierto et al., 2007b, evaluated the influence of some climate variables on pregnancy rate. Table 2 shows the final results of logistic regression. No significant effects of climatic factors on fertility were found at 7-4 and 2-1 days before AI and 1-3 days post insemination. However, it is noteworthy that for THI values > 80 three days before AI, fertility decreased from 30-34% to 23%. Although it is thought that THI > 72 units causes heat stress in cows (Johnson et al., 1962), in this study the negative effects of heat stress on fertility appeared with THI higher than 80 units.

The incidence of ovarian dysfunction in high-producing dairy cows has increased together with milk production (Wiltbank et al. 2002; López-Gatius 2003). In a recent study, Yaniz et al. (2008) showed the evolution between 1991 and 2007 of the pregnancy and ovarian disorder rates during the warm and cool seasons (Table 3). Ovarian disorders were classified as ovarian inactivity or hypofunction, cystic ovarian disease, and sub-oestrus or silent ovulation. A cow was considered to suffer ovarian hypofunction when a follicular structure of at least 8–15 mm was detected in two consecutive examinations, in the absence of a corpus luteum or cyst and no oestrus signs were noted during the 7-day period between the examinations (López-Gatius et al. 2001). Ovarian hypofunction had a clear seasonal distribution characterised by a higher frequency during the warm season (May to September) in the three five-year groupings within the study period (1991-1995, 1996-2000, and 2003-2007). However, distribution of ovarian cysts between seasons has changed over time. During the nineties, frequency of ovarian cysts was around five times higher during the warm period than during the cool period, whereas in the last five years similar proportions were recorded in both seasons (Table 3). The recent introduction of new cooler systems (fans and water sprinklers) on the farms could explain these differences.

From 2003, cows with a corpus luteum estimated to be at least 15 mm in diameter and no signs of oestrus detected by pedometers during the previous 21 days were recorded as sub-oestrous cows. The incidence of sub-oestrus cows was higher during the warm period (Table 3). An extended period of high ambient temperature reduces the intensity of oestrous behaviour (Orihuela 2000; De Rensis and Scaramuzzi 2003), and the proportion of ovulations without oestrus signs (Rodtian et al. 1996). It has been demonstrated that heat stress reduces oestradiol production by the dominant follicle (Badinga et al. 1993; Wolfenson et al. 1997).

Table 3. Seasonal effects on the pregnancy and ovarian disorder rates of cows first inseminated or examined 45-80 days postpartum (mean yearly values for the five-year periods)

Year period	1991-1995		1996-2000		2003-2007	
Season[a]	Warm	Cool	Warm	Cool	Warm	Cool
Total number of cows	2960	3346	3326	3079	3912	6581
Inseminated cows (%)	79.5	93.2	68.3	93.5	82.3	81.9
Pregnancy at first AI (%)	31.6	44.8	22.6	44.2	32.5	31.9
Ovarian hypofunction (%)	6.8	0.9	18.3	1.4	22.0	6.0
Ovarian cysts (%)	13.2	2.8	12	2.4	6.1	6.5
Sub-oestrus (%)					45.2	40.3
Milk production (kg/day/milking cow)	26.3	33.2	31.8	36.4	34.3	34.4

[a]Warm (May-September), Cool (October-April).

We can conclude that it is important to install efficient cooler systems in the farms of warm countries controlled by temperature or THI. Effectiveness of different strategies has been investigated, including evaporative cooling (Stott et al., 1972), air conditioning (Thatcher et al., 1974), installation of shades (Roman-Ponce et al., 1977), fans and water sprinklers (Wolfenson et al., 1988), and several systems combined (Ryan et al., 1992). In all the studies cited, the cooling during the summer improved the fertility, but cooling systems generally do not eliminate the seasonal depression of fertility, mainly in areas where heat stress is severe (Hansen, 1997: Collier et al., 2006).

3. Effect of Climate Factors on Pregnancy Loss

Once the cow is pregnant, it is also of great interest to account for the influence of pregnancy loss on the overall reproductive efficiency. Prenatal loss is one of the most single important factors affecting reproductive performance in high producing dairy cattle and has a substantial impact on the profitability of cow production (Ball, 1997; Vanrose et al., 2000)

The gestation of the cow has been divided into an embryonic period that extends from conception to the end of the differentiation stage (about 42 days), and a foetal period that runs from day 42 to parturition (Committee on Bovine Reproductive Nomenclature. 1972). Although most pregnancy losses occur during the early embryonic period, with a peak on days 14-16 (Peters, 1993; Inskeep and Dailey. 2005), they are not detected because most pregnancy diagnoses are usually routinely performed on days 30-40. The second peak of pregnancy losses occurs between days 45 and 60 (López-Gatius et al., 2002). These losses usually account for 10-12 % (Santos et al. 2004; BonDurant. 2007), although rates are highly variable, and early foetal loss rates of more than 20 % have been reported (Cartmill et al. 2001a; Bartolome et al. 2005a; Bartolome et al. 2005b; Bartolome et al. 2005c; Grimard et al. 2006) or even 40% (Cartmill et al. 2001b). Early foetal loss often escapes clinical diagnosis, and is assumed to be of multi-factorial origin (Lee et al. 2007). The risk of non-infectious early foetal loss appears to increase under the conditions of intensive management systems (Forar et al. 1995; Santos et al. 2004). Management- and cow-related factors of non-infectious nature, such as parity, semen-providing bull, warm season and twin pregnancies, have often been described as affecting late embryo/early foetal loss in high producing dairy cows, both in northeast of Spain (Labèrnia et al. 1996; Lopez-Gatius et al.2002; Lopez-Gatius et al., 2004c) and in other regions (Santos et al. 2004; Grimard et al. 2006). Finally, towards the end of gestation, pregnancy loss has mostly been associated with infectious diseases.

a) Early Foetal Loss

A significant correlation between season of insemination and early foetal loss was found (López-Gatius et al. 2002; López-Gatius et al. 2004a; López-Gatius et al. 2004b). For example, based on the odds ratio, cows that became pregnant during the warm period, either bearing singletons or twins, were 3.7 and 5.4 times more likely to miscarry, respectively, in comparison to the cool period (López-Gatius et al. 2002). In a more detailed and recent study, an acute heat stress during the perimplantational period predisposes pregnant cows to subsequent early foetal loss, in addition to effect of season, lactation number, twin pregnancy and additional corpus luteum. In this sense, a relationship between maximum THI on days 21-30 of gestation and subsequent foetal loss was found. Animals exposed to maximum THI values higher than 75 units during days 21-30 of gestation surpassed 30% of subsequent pregnancy loss (García-Ispierto et al. 2006).

It is well documented that heat stress can cause a reduction in dry matter intake (Hansen, 1997; Wilson et al, 1998; Drew, 1999) and may influence the uterine environment by decreasing blood flow to the uterus and increasing uterine temperature (Gwazdauskas et al., 1975;Roman-Ponce et al., 1978). Metabolic demands placed upon high-yielding dairy cows should also be added to the metabolic stress related to the warm period and could explain the manner in which season remains as a negative factor compromising the success of gestation.

During the warm season, numerous management factors could be potentially stressful for the conceptus from conception to early foetal development. These prompt the need to place emphasis on improving management practices to reduce effects of heat stress. In this sense, when stressful thermal conditions during the warm months were overcome by forced-air movement and evaporative cooling, warm season was not a risk factor for early foetal loss (Santolaria et al., 2010). The incidence of foetal loss was practically the same for cool (6.5%) and warm periods (7.1%). Only extremely hot environmental conditions could be related to foetal loss. A one-unit increase in the cumulative number of hours with maximum THI > 85 (temperatures higher than 37°C with a RH higher than 40%) on days 10-20 of the gestation period caused a 1.57-fold increase in pregnancy loss (Figure 1) (Santolaria et al., 2010). The management practices necessary for high production probably benefit from the favourable effects of fans, water sprinklers and shade when THI values are lower than 85.

Moreover, when episodes of acute increase in maximum THI values during days 0-40 of gestation were recorded, the likelihood of foetal loss increased by a factor of 1.16 for each additional episode (Figure 2). This implies that in order to maintain gestation during the late embryo/early foetal period, acute episodes of climate value changes are important. In northeast Spain, these episodes of acute increases in maximum values on the temperature-humidity index were only recorded during the cool period. In a recent study, Serrano et al., analysed conceptus-maternal interactions by measuring plasma pregnancy-associated glycoprotein (PAGs) concentrations; the development of a healthy placenta was strongly linked to gestations occurring from December to February (López-Gatius et al. 2007). Plasma PAGs levels have been used as markers of placental/foetal well-being and conception during the remaining months (March to November) was related to decreasing PAGs plasma concentrations throughout the gestation period (Serrano et al. 2009). With the above results in mind, as episodes of maximum THI increase were only recorded during the December-February period, the fact that these episodes occurring during the embryonic period were related to subsequent foetal loss is especially relevant. Further studies must be undertaken in order to understand the effects of this type of sporadic cases of heat stress during the cool period on reproductive functions in the cow. Clearly, efforts to provide appropriate environmental conditions for cow comfort should prevent the effect of sudden changes in THI during the cool season.

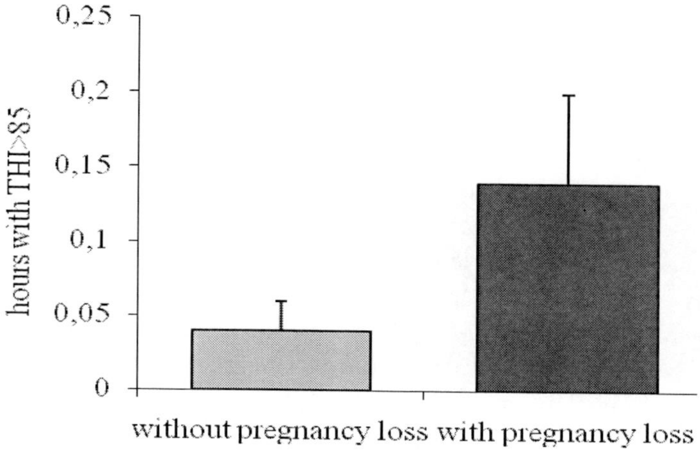

Figure 1. Number of hours with THI>85 between 10-20 days of gestation in cows with and without pregnancy loss

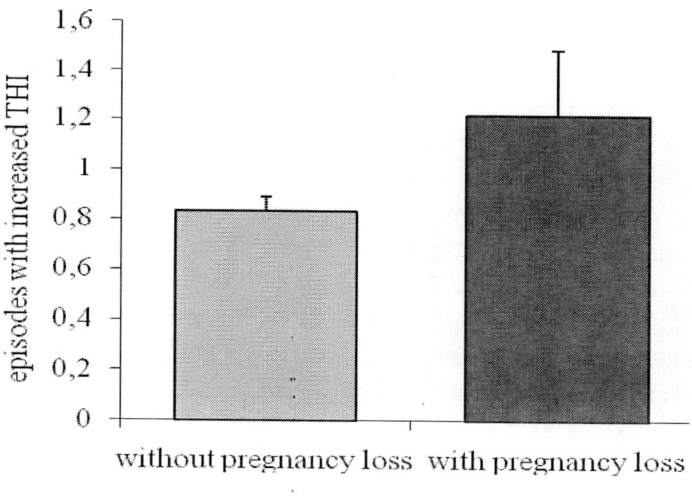

Figure 2. Number of THI oscillations (mean ± s.e.m) between 0-40 days of gestation in cows with and without pregnancy loss

During the cool season, increasing metabolic demands can also be stress-related. In a cold environment, animals react by increasing metabolic heat production, cardiac output, and hepatic gluconeogenesis and mobilisation of fat and glycogen stores (Horwitz, 1971; Girardier and Stock, 1983). One recent work, (Chebel et al. 2007) reported that conception rate of dairy heifers at 40 days after first AI was not related to exposure to cold stress between 15 days prior to and 15 days after first insemination. However, heifers exposed to cold stress were more likely to lose pregnancy between 40 and 90 days of gestation, which resulted in reduced conception rates at 90 days after first AI.

b) Late Pregnancy Loss

In contraposition with early pregnancy losses, abortion during late gestation in the cow has widely been associated with infections. Viral, bacterial, protozoan and mycoplasma infections can result in embryonic and foetal mortality, indirectly by systemic effects on the dam, or directly by affecting the embryo or contaminating its environment. Among those agents, Neospora caninum has been recognised as a major cause of abortion in cattle worldwide (Dubey et al. 2003). Neospora caninum is an obligate intracellular protozoan parasite (Bjerkås et al. 1984) that penetrates the placenta and infects the foetus causing abortion or congenital infection. Cows of any age may abort from 3 months of gestation to term, with most abortions occurring in the 5th to 7th month of gestation (Dubey et al., 2006).

In dairy cows chronically infected with N. caninum, climatic factors have been related with abortion. López-Gatius et al., 2005 reported that rainfall during the second trimester of gestation was associated with increased incidence of abortions in parous cows. Although mean rainfall in the area was very low (173 mm year) the likelihood of abortion increased progressively from 18 to 41%, for the respective rainfall mm values from <30 to 60 mm during the second trimester of gestation. In another study, Yaniz, et al. 2010 carried on with parous cows and heifers infected with N. caninum, cumulative days with mean RH lower than 60% during the second trimester of gestation were associated with the risk of abortion. Similarly to the previous study, it was confirmed that rainfall increased the risk of abortion in parous cows. However, rainfall had no effect on heifers. The difference regarding rainfall could not be attributed to the housing conditions, as parous cows were housed in free stalls under cover, and consequently more protected against the effect of rainfall than heifers, housed in free stalls in outside lots. Nor could it be attributed to a higher sensitivity of parous cows to N. caninum infection, as the incidence of abortions was slightly lower in cows (21.9 %) than in heifers (27.5 %). Heifers probably suffer lower metabolic stress related to milk production and are able to support environmental stresses better than lactating cows.

The mechanism whereby cumulative days with low RH and rainfall during the second trimester of gestation increased the incidence of abortion is uncertain, but might involve a reduction in the already down-regulated cell-mediated immune response at mid-gestation (Innes et al., 2001). It seems that the activity of the immune system varies depending on environmental conditions. For example, seasonal alterations in the function of the immune system have been described in mammalian populations (Nelson et al., 1995; Nelson, 2004; Mann et al., 2000; Freeman, 2008). Although seasonal immunity seems to be modulated by changes in day length, it is possible that other environmental conditions such as the relative humidity and rainfall may also have some influence in the immune function, changing the susceptibility to disease. For example, rainfall may be associated with the development of mouldy fodder, which feeding is considered to be a factor which may induce recrudescence of a latent N. caninum-infection by mycotoxins though the induction of immune suppression (Bartels et al., 1999).

REFERENCES

Badinga, L; Thatcher, WW; Diaz, T; Drost, M; Wolfenson, D. Effect of environmental heat stress on follicular development and stereidogenesis in lactating dairy cows. *Theriogenology*, 1993, 39, 797-810.

Ball, PJH. Later embryo and early fetal mortality in the cow. *Animal Breeding Abstract*, 1997, 65, 167-75.

Bartels, CJM; Wouda, W; Schukken, YH. Risk factors for Neospora caninum-associated abortion storms in dairy herds in the Netherlands (1995 to 1997). *Theriogenology*, 1999, 52, 247-257

Bartolome, JA; Silvestre, FT; Kamimura, S; Arteche, AC; Melendez, P; Kelbert, D; McHale, J; Swift, K; Archbald, LF; Thatcher, WW. Resynchronization of ovulation and timed insemination in lactating dairy cows I: use of the Ovsynch and Heatsynch protocols after non-pregnancy diagnosis by ultrasonography. *Theriogenology*, 2005a, 63, 1617-1627.

Bartolome, JA; Sozzi, A; McHale, J; Melendez, P; Arteche, AC; Silvestre, FT; Kelbert D; Swift, K; Archbald, LF; Thatcher, WW. Resynchronization of ovulation and timed insemination in lactating dairy cows, II: assigning protocols according to stages of the estrous cycle, or presence of ovarian cysts or anestrus. *Theriogenology*, 2005b, 63, 1628-1642.

Bartolome, JA; Sozzi, A; McHale, J; Swift, K; Kelbert, D; Archbald, LF; Thatcher, WW. Resynchronization of ovulation and timed insemination in lactating dairy cows III. Administration of GnRH 23 days post AI and ultrasonography for nonpregnancy diagnosis on day 30. *Theriogenology*, 2005c, 63, 1643-1658.

Beede, DK; Shearer, JK. Thermoregulation and Physiological Responses of Dairy Cattle in Hot Weather. *Agri- Practice*, Vol. 11, No. 4. July/August. 1991.

Berman A. Estimates of heat stress relief needs for Holstein dairy cows. *Journal of Animal Science*, 2005, 83, 1377-1384.

Bjerkås, I; Mohn, SF; Presthus, J. Unidentified cyst-forming sporozoon causing encephalomyelitis and myostitis in dogs. *Z. Parasitenkd*, 1984, 70, 271-274.

BonDurant, RH. Selected diseases and conditions associated with bovine conceptus loss in the first trimester. *Theriogenology*, 2007, 68, 461-473.

Broucek, J; Letkovicová, M; Kovalcuj, K. Estimation of cold stress effect on dairy cows. *International Journal of Biometeorology*, 1991, 35, 21-32

Bufimgton, DE; Colazo-Amcho, A; Canton, GH; Pia, D; Thatcher, WW; Wilcox, RJ. Black-globe-humidity-index (BGHI as comfort equation for dairy cows). *Trans Am Soc Agri Eng.*, 1981, 711

Cartmill, JA; El-Zarkouny, SZ; Hensley, BA; Lamb, GC; Stevenson, JS. Stage of cycle, incidence, and timing of ovulation, and pregnancy rates in dairy cattle after three timed breeding protocols. *Journal of Dairy Science*, 2001a, *84*, 1051-1059.

Cartmill, JA; El-Zarkouny, SZ; Hensley, BA; Rozell, TG; Smith, JF; Stevenson, JS. An alternative AI breeding protocol for dairy cows exposed to elevated ambient temperatures before or after calving or both. *Journal od Dairy Science*, 2001b, 84, 799-806.

Chebel, RC; Braga, FA; Dalton, JC. Factors affecting reproductive performance. *Animal Reproduction Science*, 2007, 101, 208-224.

Collier, RJ; Dahl, GE; VanBaate, MJ. Major advances associated with environmental effects on dairy cattle. *Journal of Dairy Science*, 2006, 89, 1244-1253

Committee on Bovine Reproductive Nomenclature, Recommendations for standardizing bovine reproductive terms. *Cornell Veterinary*, 1972, 62, 216-37.

De Rensis, F; Scaramuzzi, RJ. Heat stress and seasonal effects on reproduction in the dairy cow-a review. *Theriogenology*, 2003, 60, 1139-1151.

Dobson, H; Smith, R; Royal, M; Knight, C; Sheldon, I. The High-producing Dairy Cow and its Reproductive Performance. *Reproduction in Domestic Animals*, 2007, 42 Suppl 2 17-23.

Drew, B. Practical nutrition and management of heifers and high yielding dairy cows for optimal fertility. *Cattle Practice*, 1999, 7, 243-8.

Dubey, JP. Review of *Neospora caninum* and neosporosis in animals. *Korean Journal of Parasitology*, 2003, 41, 1-16.

Dubey, JP; Buxton, D; Wouda, W. Pathogenesis of bovine neosporosis. *Journal Comp Pathology*, 2006, 134, 267-289.

Erb, RE; Wilbur, JW; Hilton, JH. Some factors affecting breeding efficiency in dairy cattle. *Journal of Dairy Science*, 1940, 23, 549

Folk, GE. Textbook of environmental physiology. Lea & Febiger, *Philadelphia, Febiger*, 1974, 97-118

Forar, AL; Gay, JM; Hancock, DD. The frequency of endemic fetal loss in dairy cattle: A review. *Theriogenology*, 1995, 43, 989-1000.

Freeman, DA. Multiple neuroendocrine pathways mediate seasonal immunity. *Am. J. Physiol. Regul. Integr. Comp. Physiol.*, 2008, 294 382-383.

Fuquay, JW. Heat stress as it affects animal production. *Journal of Animal Science*, 1981, 52, 16.

García-Ispierto, I; López-Gatius, F; Bech-Sabat, G; Santolaria, P; Yániz, JL; Nogareda, C; De Rensis, F; López-Béjar, M. Climate factors affecting the fertility of high producing dairy cows in northeastern Spain. *Theriogenology*, 2007b, 67, 1379-1385.

García-Ispierto, I; López-Gatius, F; Santolaria, P; Yániz, JL; Nogareda, C; López-Béjar, M. Factors affecting the fertility of high producing dairy herds in northeastern Spain. *Theriogenology*, 2007a, 67, 632-638.

García-Ispierto, I; López-Gatius, F; Santolaria, P; Yániz, JL; Nogareda, C; López-Béjar M; De Rensis, F. Relationship between heat stress during the peri-implantation period and early fetal loss in dairy cattle. *Theriogenology*, 2006, 65, 799-807.

Gaughan, JB; Mader, TL; Holt, SM; Lisle, A. A new heat load index for feedlot cattle. *Journal of Animal Science*, 2008, 86 (1), 226-234

Girardier, L; Stock, MJ. *Mammalian Thermogenesis*. Chapman and Hall, University Press, Cambridge, 1983.

Grimard, B; Freret, S; Chevallier, A; Pinto, A; Ponsart, C; Humblot, P. Genetic and environmental factors influencing first service conception rate and late embryonic/foetal mortalitiy in low fertility dairy herds. *Animal Reproduction of Science*, 2006, 91, 31-44

Gwazdauskas, FC; Wilcox, CJ; Thatcher, WW. Environmental and management factors affecting conception rate in a subtropical climate. *Journal of Dairy Science*, 1975, 58, 88-92.

Hahn, GL. Predicted vs. measured production differences using summer air conditioning for lactating cows. *Journal of Dairy Science*, 1969, 52, 800-801.

Hansen, PJ. Effects of environment on bovine reproduction. In: Youngquist RS., editor. *Current therapy in large animal theriogenology*. WB Saunders Company, 1997, 403-415.

Hansen, PJ. Embryonic mortality in cattle from the embryo's perspective. *Journal of Animal Science*, 2002, 80(2), 33-44.

Horwitz, BA. Brown fat thermogenesis: physiological control and metabolic basis. Nonshivering thermogenesis. *Proceedings of Symposium*, Praha, 1971, 63-65.

Innes, EA; Wright, SE; Maley, S; Rae, A; Schock, A; Kirvar, E; Bartley, P; Hamilton, C; Carey, IM; Buxton, D. Protection against vertical transmission in bovine neosporosis. *International Journal of Parasitology*, 2001, 31, 1523-1534.

Inskeep, EK; Dailey, RA. Embryonic death in cattle. *Veterinary Clinics of North America-Food Animal Practice*, 2005, 21, 437.

Johnson, HD; Ragsdale, AC; Berry, IL; Shanklin, MD. Effect of various temperature–humidity combinations on milk production of Holstein cattle. *Missour Agric Exp Stn Res Bull*, 1962, 791, 1-39.

Labèrnia, J; López-Gatius, F; Santolaria, P; Hanzen, C; Laurent, Y; Houtain, JY. Influence of calving season on the interactions among reproductive disorders of dairy cows. *Animal Science*, 1998, 167, 387-393.

Labèrnia, J; López-Gatius, F; Santolaria, P; López-Béjar, M; Rutllant, J. Influence of management factors on pregnancy attrition in dairy cattle. *Theriogenology*, 1996, 45, 1247-1253.

Lee, JI; Kim, IH. Pregnancy loss in dairy cows: the contributing factors, the effects on reproductive performance *Journal Veterinary Science*, 2007, 8(3), 283-8.

López-Gatius, F. Is fertility declining in dairy cattle? A retrospective study in northeastern Spain. *Theriogenology*, 2003, 60, 89-99.

López-Gatius, F; Garbayo, JM; Santolaria, P; Yániz, JL; Almeria, S; Ayad, A; de Sousa, NM; Beckers, JF. Plasma pregnancy-associated glycoprotein-1 (PAG-1) concentrations during gestation in Neospora-infected dairy cows. *Theriogenology*, 2007, 67, 502-508.

López-Gatius, F; Pabón, M; Almería, S. Neospora caninum infection does not affect early pregnancy in dairy cattle. *Theriogenology*, 2004c, 62, 606-613.

López-Gatius, F; Santolaria, P; Yaniz, J; Rutllant, J; López-Béjar, M. Factors affecting pregnancy loss from gestation day 38 to 90 in lactating dairy cows from a single herd. *Theriogenology*, 2002, 57, 1251-1261.

López-Gatius, F; Santolaria, P; Yániz, J; Rutllant, J; López-Béjar, M. Persistent ovarian follicles in dairy cows: a therapeutic approach. *Theriogenology*, 2001, 56, 649-659.

López-Gatius, F; Santolaria, P; Yániz, JL; Garbayo, JM; Hunter, RHF. Timing of early foetal loss for single and twin pregnancies in dairy cattle. *Reproduction in Domestic Animals*, 2004a, 39, 429-433.

López-Gatius, F; Santolaria P; Yániz JL; Hunter RHF. Progesterone supplementation during the early fetal period reduces pregnancy loss in high-yielding dairy cattle. *Theriogenology*, 2004b, 62, 1529-1535.

López-Gatius, 2005.

Lucy, MC. Reproductive Loss in High-Producing Dairy Cattle: Where Will It End?. *Journal of Dairy Science*, 2001, 84, 1277-1293.

Lucy, MC. Fertility in high-producing dairy cows: reasons for decline and corrective strategies for sustainable improvement. *Soc. Reprod. Fertil. Suppl*, 2007, 64, 237-254.

Mann, DR; Akinbami, MA; Gould, KG; Ansari, AA. Seasonal variations in cytokine expression and cell-mediated immunity in male rhesus monkeys. *Cell Immunology*, 2000, 200, 105-115.

Mc Dowell, D; Hooven N; Cameron, K. Effects of climate on performance of Holsteins in first lactation. *Journal of Dairy Science*, 1979, 68, 2418-35.

Mc Dowell, RE. *Improvement of livestock production in warm climate*. Freeman, San Francisco, 1972, 410-449.

Nelson, RJ. Seasonal immune function and sickness responses. *Trends in Immunology*, 2004, 25, 187-192.

Nelson, RJ; Demas, GE; Klein, SL; Kriegsfeld, LJ. The influence of season, photoperiod, and pineal melatonin on immune function. *Journal of Pineal Research*, 1995, 19, 149 -165.

Orihuela, A. Some factors affecting the behavioural manifestation of oestrus in cattle: a review. *Applied of Animal Behaviour Science*, 2000, 70, 1-16.

Perry, JS; Rowlands, IW. Early pregnancy in the pig. *Journal of Reproduction Fertility*, 1962, 4, 175-88.

Peters, AR. Embryo mortality in the cow. *Animal Breeding Abstract*, 1996, 64, 587-98.

Reynolds, LP; Redmer, DA. Growth and microvascular development of the uterus during early pregnancy in ewes. *Biology Reproduction*, 1992, 47, 698-708.

Rodtian, P; King, G; Subrod, S; Pongpiachan, P. Oestrous behaviour of Holstein cows during cooler and hotter tropical seasons. *Animal Reproduction Science*, 1996, 45, 47-58.

Roman-Ponce, H; Thatcher, WW; Canton, D; Barron, DH; Wilcow, CJ. Thermal stress effects on uterine blood flow in dairy cows. *Journal of Animal Science*, 1978, 46, 175-80.

Roman-Ponce, H; Thatcher, WW; Buffington, DE; Wilcox, CJ; Van Horn, HH. Physiological and Production Responses of Dairy Cattle to a Shade Structure in a Subtropical Environment. *Journal of Dairy Science*, 1977, 60(3), 424-430.

Ryan, DP; Boland, MP; Kopel, E; Armstrong, D; Munyakazi, L; Godke, RA; Ingraham, RH. Evaluating Two Different Evaporative Cooling Management Systems for Dairy Cows in a Hot, Dry Climate. *Journal of Dairy Science*, 1992, 75(4) 1052-1059.

Santolaria, P; Lopez-Gatius, F; Garcia-Ispierto, I; Bech-Sabat, G; Angulo, E; Carretero, T; Sánchez-Nadal, JA; Yániz, J. Effects of cumulative stressful and acute variation episodes of farm climate Inter. *Journal of Biometeorology*, 2010, 54, 93-98

Santos, JEP; Thatcher, WW; Chebel, RC; Cerri, RLA; Galvão, KN. The effect of embryonic death rates in cattle on the efficacy of estrus synchronization programs. *Animal Reproduction Science*, 2004, 82, 513-535

Seath, DM; Staples, CH. Some factors influencing the reproductive efficiency in Lousiana herds. *Journal of Dairy Science*, 1941, 24, 510.

Serrano, B; López-Gatius, F; Santolaria, P; Almeria, S; García-Ispierto, I; Bech-Sàbat, G; Sulon, J; de Sousa, NM; Beckers, JF; Yániz, JL. Factors affecting plasma pregnancy-associated glycoprotein concentrations throughout gestation in high producing dairy cows. *Reproduction in Domestic Animals*, 2008, 44, 600-605.

Stott, GH; Wiersma, F; Woods, JM. Reproductive health program for cattle subjected to high environmental temperatures. *Journal os the American Veterinary Medical Association*, 1972, 161(11), 1339-1344.

Thatcher, WW; Meyer, MD; Daret-Desoroyers, G. Maternal recognition of pregnancy. *Journal Reproduction Fertility*, 1995, 49,15-28.

Thom, EC. Cooling degrees-days. *Air Cond. Heating Ventil*, 1958, 53, 65-72.

Thom, EC. The discomfort index. *Weatherwise*, 1959, 12, 57-60.

Vanroose, G; de Kruif, A; Van Soom, A. Embryonic mortality and embryo-pathogen interactions. *Animal Reproduction Science*, 2000, 60-61, 131-43.

Thatcher, WW; Gwazdauskas, FC; Wilcox, CJ; Toms, J; Head, HH; Buffington, DE; Fredricksson, WB. Milking Performance and Reproductive Efficiency. *Journal of Dairy Science*, 1974, 57(3), 304-307.

Wiersama, F. Dep. Agric. Eng., University of Arizona, Tucson, Arizona, USA. 1990.

Wilson, SJ; Marion, RS; Spain, JN; Spiers, DE; Keisler, DH; Lucy, MC. Effect of controlled heat stress on ovarian function in dairy cattle: I Lactating cows. *Journal of Dairy Science*, 1998, 1, 2124-31.

Wiltbank, MC; Gumen, A; Sartori, R. *Physiological classification of anovulatory conditions in cattle theriogenology*, 2002, 57, 21-52.

Wolfenson, D; Lew, BJ; Thatcher, WW; Graber, Y; Meidan, R. Seasonal and acute heat stress effects on steroid production by dominant follicles in cows. *Animal Reproduction Science*, 1997, 47, 9-19.

Wolfenson, D; Flamenbaum, I; Folman, Y; Kaim, M; Berman, A. Thermal, Productive, and Reproductive Responses of High Yielding Cows Exposed to Short-Term Cooling in Summer. *Journal of Dairy Science*, 1988, 71(4), 1085-1092.

Yániz, JL; Lopez-Gatius, F; Bech-Sàbat, G; García-Ispierto, I; Serrano, B; Santolaria, P. Relationship between milk production, ovarian function and fertility in high-producing dairy herds in north-eastern Spain. *Reproduction in Domestic animals*, 2008, 43, 38-43.

Yániz, JL; López-Gatius, F; García-Ispierto, I; Bech-Sàbat, G; Serrano B; Nogareda C; Sanchez-Nadal, JA; Almeria, S; Santolaria P. Some factors affecting the abortion rate in dairy herds with high incidence of *neospora*-associated abortions are different in cows and heifers. *Reproduction in Domestic Animals*, in press. 2010.

Yousef, MK. In: Basic Principles. Stress Physiology in Livestock. Vol 1. CRC Press, *Boca Raton*, FL. 1985.

In: Relative Humidity: Sensors, Management…
Editor: John M. Castillo

ISBN: 978-1-61761-734-8

Chapter 6

MOLDS IN BUILDING MATERIALS OF AGRA (INDIA) AT VARIOUS HUMIDITY CONDITIONS

Alka Jain[*]
Ch. Raghunath Singh Mahavidyalaya,
Kiraoli Agra, R.B.S. College Agra, India

ABSTRACT

Three parameters relative humidity, temperature and organic substances are essential for the growth of molds. Building material like sandstone, marble, granite etc. are the inorganic material on which the growth of molds and other microbes is not favored, It was likely that organic residues from different sources on stone surface were utilized as nutrients. Light, temperature, humidity combined with rich nutrients of organic residues present on stone surface create suitable conditions for reproduction of dormant mold spores.

To study the effect of relative humidity – molds were isolated from three monuments Agra fort, Christian Cemetry and Kailash temple in three seasons (summer, rain and winter with relative humidity 50–60%, 80-98%, 70-85% respectively). The data showed maximum fungal diversity in rain season followed by winter season. In summer season, thermophilic fungi appeared in the procedure of isolation. Besides the deteriorated sandstone samples were also used for direct isolation of molds.

The effect of relative humidity was also confirmed by using tropical chamber test. This test used saturated chemicals to maintain definite relative humidity (like 96.6%, 85%, 76%). During the experimental study scanning electron microscopy was used for the analysis of pattern of biodeterioration. Many sporulating fungi and filamentous fungi were found to play active role in diagenesis of stone surface. A network of fungal hyphae penetrated the sandstone substrate and detached the loose arranged sand grains. The fungi bioweatherd silicates and other minerals through its biochemical and biomechanical activity. Besides, some crystals have also been found to be attached on their surface, which has been shown by SEM and XRD.

---

[*] Corresponding author: e-mail- alka.jain@rediffmail.com

Introduction

Building material like sandstone is readily colonized by virtue of their porous nature and mineral composition. Interaction with the atmosphere may cause deposition of organic particulate matter, containing microorganisms. Deterioration process may be triggered by exogenous factors for example humidity, poor ventilation and light (including artificial light) facilitate formation of biofilms composed of viscous EPS (extra polymeric substance), which are hydrogels consisting of 98% of water that adhere to the surface. Hydration and formation of biofilms that collect particles and atmospheric contaminants accelerate chemical corrosion (Warscheid and Braams 2000; Beech 2004) with oxidation, reduction and transformation of metal ions and pigment changes (Urzi and Realni 1998). Uncontrolled development of this phenomenon often coexists with saline efflorescences, also promoted by high humidity (Gorbushina et al. 2004). The presence of salts may influence the properties of EPS by decreasing their viscosity and causing surface irregularities which are preferential sites for colonization by various communities of microorganisms (Roldan et al. 2003) and make stone bioreceptive (Giullitte, 1995), increasing microbial concentration and diversification.

Molds are considered one of the most efficient groups of microorganisms causing biodeterioration of stone monuments. They contribute to a great variety of alterations like aesthetical spoiling of the monuments, due to colour changes and black spots, intergranular growth, biopitting etc. (Urzi et al. 2000).

The ability of fungi to colonize the substrate is greatly enhanced by different factors among which the most important are the environmental factors, nutrients and humidity. During the investigations two main groups of fungi are usually isolated from rock surfaces.

One group includes species of genera of Hyphomycetes and Coelomycetes. Either they produce melanin or not. Most members of these group are ubiquitous (Domsch 1980) and widely distributed. They normally occur as saprophytes in soil and characterized by fast growth in normal cultural media, relatively high water activity and eutrophic nutrient conditions (Sterflinger 2000).

The second group includes the so-called black Yeast and meristematic fungi. They represent a wide and heterogenous group of black pigmented fungi that share as common characteristics – the presence of melanin within the cells (swollen cells, hyphae and spores). The production of melanin and the meristematic development allow them to survive in stressed environmental conditions like low relative humidity, high temperature, high incidence of UV-light and low nutrients. The fungi of this group is also termed as ‘rock inhibiting fungi’ or ‘microcolonial fungi (MCF)’ to underline the excessive isolation of many of them from rock surfaces. MCF contains melanin, mycosporines, carotenoids and possibly other unidentified pigments that help to protect them from UV-light. Vegetative cells of these fungi have complex intracellular division pattern or a yeast like budding pattern. Vegetative cells of the microcolonies of MCF are highly stress tolerant and extremely long living. They found on rock and rock coatings of arid and semiarid deserts (Perry et al. 2003).

In many cases it was observed that fungus are the first colonizers of stone surface (Urzi et al. 2000). Melanized fungi in particular, are the most common species isolated and many of them present structures that allow them to survive in dry conditions or with a low requirement of nutrients and energy and to be less susceptible to seasonal variation facilitating the colonization of stone surfaces.

Excessive moisture supports the microbial growth. The relative humidity more than 55% creates the niche that is required for the germination of quiescent (dormant) fungal conidia and other spores. These molds react to stones by chemical and mechanical action and cause physical and chemical alterations. On the other hand the surface water near the stone surface combined with CO2 from microbial respiration and produced carbonic acid that could dissolve the Ca and Mg carbonates, the constituents of stones. The dissolution of constituents of stone caused changes of stone composition and material loss. Besides, the bioweathering of mineral substrate results in the formation of secondary biominerals on the attacked substrates.

This study was performed to evaluate the change in diversity of molds during the seasonal fluctuations of relative humidity; this would also studied under in-vitro condition of plant growth chamber test. Scanning electron microscopic study (SEM) is and important tool that was used for the analysis of procedure of bioweathering, aliet the formation of new minerals has also been investigated successfully.

MATERIALS AND METHODS

Areas of sampling – Three sandstone monuments were selected for the experimental study. These three monuments belonged to three categories:-

- Sampling site under world heritage – Agra fort, Agra
- Sampling site not listed in world heritage whereas conserved by Archaeological survey of India – European or Christian Cemetry Agra
- Selected sampling site left unconserved however is of religious importance – Kailash temple Agra

Agra fort – This fort was basically built by the great Mughal Emperor Akbar. This fort was constructed during 1565 A.D. to 1573 A.D. The construction was entrusted to Kashim Khan who was commander-in-chief of Akbar and Governor of Kabul. The total sum spent on the construction was Rs. 3.5 millions at that time. This huge and beautiful fort is an amalgamation of buildings and palaces. Not only Akbar but also his son Jahangir, grandson Shah Jahan and great-grandson Aurangzeb time to time contributed for various buildings, mosque and palaces. Therefore it is vast museum of buildings of various tastes and nature of Mughal rulers.

Agra fort

Agra fort is situated at the right bank of river Yamuna towards the east of Agra city. It has been built of thick and strong walls of red sandstone. Only the buildings of Akbar's period are of sandstone. Other beautiful palaces and mosque are of marble. The architecture of fort is famous for its blend of hindu and central Asian styles – a contrast to the typical Mughal style specially Jahangir palace.

European or Christian Cemetry, Agra - European Cemetry is situated on Mahatma Gandhi road, opposite side of the court of district judge Agra. It is about 300 years old graveyard of Roman Catholics. There are many old graves- the oldest grave was built in 1611 A.D. Some graves have domes and couplas over them like Mughal mausoleums. The direction of these graves is north and south. The most beautiful mausoleum is of John William Hessing which looks like mini Taj Mahal of red sandstone. John William Hessing was appointed as custodian of Agra fort in 1799 A.D. He died in 1803 at Agra.

European Cemetry

Kailash Temple - This is very old temple of God Shiva situated on the bank of river Yamuna, about 2 km away from Sikandara, the mausoleum of Mughal emperor Akbar. It is believed by temple authority that the temple is 1500 years old. In 18th century it was renovated by Marathas. At present time only pillars, stairs i.e. the architectural view in front of river is of ancient construction. Main temple has been renovated completely and is constructed of marble.

Kailash temple Agra(India)

Method of Sampling

Swabbing and Serial dilution method (Nigam, 1965): In this method the surface of deteriorated sandstone sample was swabbed by sterilized moist cotton and shaken in 10 ml of sterilized distilled water. Serial dilutions $10^{-2}, 10^{-3}$.........10^{-7} were made by pipetting measured volumes (1ml) into additional dilution blanks (having 9 ml sterile water). Finally 1 ml aliquots of various dilutions were added to sterile petri plates (triplicate for each dilution) to which were added 20 ml of the sterile, cool molten (45°C) media (Czapeck-dox agar/ rose Bengal agar for fungi). The dilutions 10^{-2} to 10^{-5} were selected for enumeration of fungi. Upon solidification, the plates were incubated at 25°C for 3-7 days.

The isolated species of fungi were identified following Gilman (2001), Barnett and Hunter (1972) and Ellis (1976).

Quantitative Determination and Comparison of Population

(A) The **percentage abundance** of the microorganisms was calculated by the following formula:

$$\text{Percent abundance} = \frac{\text{Total population of single organism}}{\text{total population of all organisms}} \times 100$$

(B) Diversity index of microorganisms

The Shannon and Weaver index is a diversity index. The higher the value of H, the greater is the diversity. The maximum value of H can be more than 1.

$$\text{Shannon diversity index } H = -\Sigma\, p_i \ln p_i \quad \text{Where} \quad pi = \frac{n_i}{N}$$

ni is the number or biomass or productivity of each mass and N is the total value for all species in the community (Dash, 1993).

(C) Relative frequency

The relative frequency (RF) is determined by the use of the following formula

$$\text{R.F. of the species} = \frac{\text{number of occurrences of a species}}{\text{Number of occurrences of all species}} \times 100$$

Direct Plating Method

In this method, small pieces of the deteriorated samples were directly placed on the surface of agar media in Petri-plates under aseptic conditions and incubated at suitable temperatures for 7 days in incubator. The developing microbes were observed.

Tropical Chamber Test

Tropical chamber method is a modified method in which desiccators are used as tropical chamber. In these desiccators relative humidity of 52%, 76%, 85% & 96.6% is maintained by using saturated solution of $CaNO_3$, NaCl, KCl, KH_2PO_4 respectively. The lids are sealed with grease to avoid loss of vapours, so that desiccators could keep at required temperature (25°C). The relative humidity inside the desiccators was regularly checked after four days using a dial hygrometer. The pieces were taken out after every 6 months upto 24 months for colonization of mycoflora. The tropical chamber with relative humidity 52% was considered as control since the growth of microbes is generally inhibited below 62%. The sampling was carried out in correspondence of the fibrous growth on stone surface by needle technique (Wollenzein et al. 1995).

SEM study - The standard method of sample preparation (Jain et al 2009) was used for scanning electron microscopy. The samples were examined with Hitachi 3000N SEM and Zeiss EVO 40 EP.

EDX: Energy dispersive X-ray analysis was done on Bruker LN2 free X-flash 4010SDD detector attached to Zeiss EVO 40 EP SEM and spectra was analysed by using QUANTAX 200 software. The samples were coated with carbon in carbon coating unit before application of this technique.

Material characterization: weathered or fungal infested sandstone samples were made characterized by X-ray diffraction analysis (Phillips – XPERT-PRO PANalytical) from Wadia Institute of Geology Dehradun, India.

Results and Discussion

Isolation of Molds from Monuments

The results of isolation have been revealed from table 1 to table 3 for three monuments viz. Agra fort, European Cemetry and Kailash temple. At the site of sampling there was black patina, exfoliation and cracks on the walls. The diversity of fungi was measured quantitatively by Shannon and Weaver index. On the sandstone of Agra fort the diversity of fungi was maximum in the season of rain when the relative humidity was in the range of 80% to 100% (H = 0.784), followed of winter (H=0.493) and summer (H=0.471). In rain season most abundant species was *Rhizoctonia solani* followed by *Bipolaris sp.* and *Aspergillus fumigatus*. Other species isolated from the site were *Monilia sp., Alternaria alternata* (plate 1 Figure a), *Aspergillus flavus, A. niger, Helicocephalum sp.* and *Mycelia sterilia*.

In the season of winter most abundant species was *Aspergillus terreus*. The fungi *Fusarium sp.* and *Penicillium sp.* only appeared during winter (R.H. = 70% to 85%), not any other season. Two species like *A. terreus* and *Papulospora sp.* (plate 1 Figure b) appeared in two season summer and winter but not in rain.

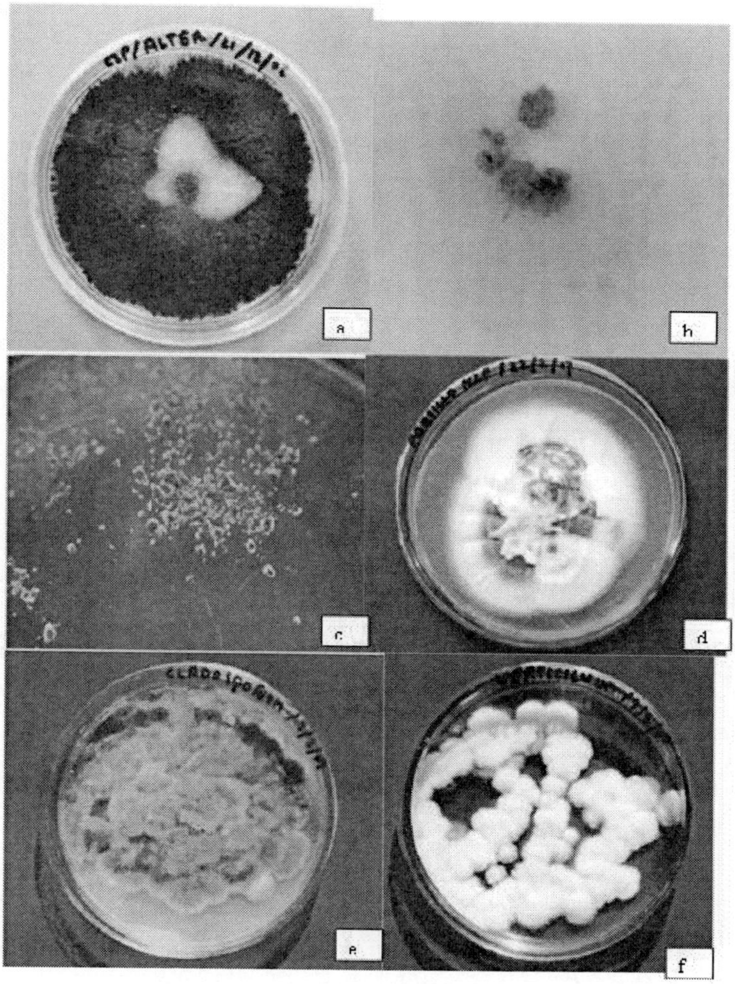

Plate 1.(a). Pure plate of *Alternaria alternata* isolated from Agra fort, Agra
(b). Pure plate of *Papulospora sp.* isolated from Agra fort, Agra
(c). Pure plate of *Trichoderma koningi* isolated from European Cemetry, Agra
(d). Pure plate of *Paecilomyces sp.* isolated from Kailash temple, Agra
(e). Pure plate of *Cladosporium cladosporoides* isolated from Kailash temple, Agra
(f). Pure plate of *Verticillium alboatrum* isolated from Kailash temple, Agra

During the season of summer when relative humidity was upto 60% species of *Aspergillus, Cunninghamella, Curvularia subulata, Candida sp., Alternaria alternata, Papulospora sp.* were isolated from the swabbed area of this monument.

European Cemetry as well as Agra fort is made up of red sandstone. European Cemetry is situated in the heart of city therefore more exposed to pollutants of city. This is reflected by more fungal species isolated form this monument (table 2). Some of the species of fungi like *Gilmaniella sp., Staphylotrichum sp., Humicola sp. Cephalosporium curtipes, Septonema sp., Aspergillus versicolor* and *Trichoderma koningi* (plate 1 Figure c) were new in comparison to the fungal flora of Agra fort. During the season of rain most abundant sp. was *Curvularia geniculata* followed by *Bipolaris sp., Aspergillus flavus*. The Shannon diversity index was 0.558 maximum among the flora for three seasons. Quality analysis of population during the summer showed that *Humicola sp., Candida sp., Papulospora sp.* and *Aspergillus flavus* were most abundant species. On the other hand during winter *Cephalosporium curtipes* was most abundant form followed by *Rhizoctonia solani, Bipolaris sp., Papulospora sp.*

Quality analysis of fungi for the monument-Kailash temple (table 3) revealed that total 17 species were isolated in three seasons by swabbing method. The variety of fungal spectrum depends mainly on seasonal variation i.e. fluctuation in relative humidity and organic matter deposited on the surface of stone. In summer the fungal flora revealed the dominance of *Aspergillus fumigatus*. Other abundant species were *A. flavus, A. niger, A. sp., Alternaria alternata, Trichoderma koningi* and *Curvularia subulata* belonging to Deuteromycetes. Only one member of Phycomycetes *Rhizopus sp.* grew in population. In rain season the abundant fungus were *A. fumigatus, Curvularia geniculata* and *Penicillium sp.* Analysis of winter season showed the abundance of *Aspergillus sydowi*, and *Paecilomyces sp.* There was the introduction of new species like *Verticillium alboatrum, Spiromyces sp.* which were not find in the isolation method from other two monuments. The pure plates of *Paecilomyces sp.*, *Cladosporium cladosporoides, V. alboatrum,* has been shown in plate 1 (Figure d, e, f respectively).

Table 1. Occurrence of mycoflora in different seasons on Agra fort, Agra

Fungi	Summer		Rain		Winter	
	Swab		Swab		Swab	
	TP	%A	TP	%A	TP	%A
Aspergillus flavus	1	9.09	1	5	-	-
A.niger	1	9.09	1	5	-	-
A.fumigatus	1	9.09	2	10	-	-
A.terreus	1	9.09	-	-	3	21.4
Cunninghamella sp.	1	9.09	-	-	-	-
Fusarium sp.	-	-	-	-	-	-
Helicocephalum sp.	-	-	1	5	-	-
Gilmaniella sp.	-	-	-	-	-	-
Rhizoctonia solani	-	-	9	45	1	7.14
Bipolaris sp.	-	-	3	15	2	14.2
Monilia sp.	-	-	1	5	1	7.14
Candida sp.	2	18.1	-	-	1	7.14
Curvularia subulata	2	18.1	-	-	-	-
Papulospora sp.	1	9.09	-	-	1	7.14
Penicillium sp.	-	-	-	-	1	7.14
Alternaria alternata	1	9.09	1	5	2	14.2
Mycelia sterilia	-	-	1	5	-	-
Fusarium sp.					2	14.2
Total	11		20		14	
Shannon diversity index	0.471		0.784		0.493	

TP =Total population Swabbing affected area size = 4x4 cm^2

%A= Percentage abundance - = absent

Table 2. Occurrence of mycoflora in different seasons on European Cementry, Agra

Fungi	Summer		Rain		Winter	
	Swab		Swab		Swab	
	TP	%A	TP	%A	TP	%A
Aspergillus flavus	3	14.2	4	11.76	-	-
A.niger	-	-	3	8.82	-	-
A.versicolor	-	-	2	5.88	1	6.66
Cephalosporium curtipes	1	4.76	1	2.94	3	20
Humicola sp.	4	19.0	-	-	-	-
Staphylotrichum sp.	-	-	2	5.88	-	-
Gilmaniella sp.	-	-	2	5.88	-	-
Rhizoctonia solani	2	9.52	1	2.94	2	13.3
Bipolaris sp.	2	9.52	4	11.76	2	13.3
Septonema sp.	-	-	-	-	1	6.66
Candida sp.	3	14.2	2	5.88	1	6.66
Curvularia geniculata	1	4.76	8	23.5	-	-
Papulospora sp.	3	14.2	2	5.88	2	13.3
Penicillium livida	-	-	2	5.88	-	-
Alternaria alternata	1	4.76	1	2.94	-	-
Mycelia sterilia	1	4.76	-	-	1	6.66
Trichoderma koningi	-	-	-	-	1	6.66
Unknown	-	-	-	-	1	6.66
Total	21		34		15	
Shannon diversity index		0.463		0.558		0.466

TP =Total population Swabbing affected area size = 4x4 cm^2
%A= Percentage abundance - = absent

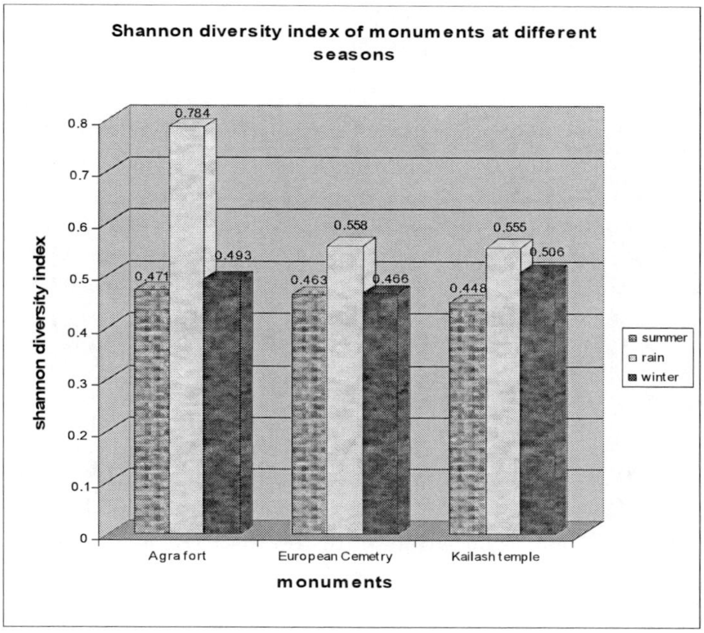

Table 3. Occurrence of mycoflora in different seasons on Kailash temple, Agra

Fungi	Summer		Rain		Winter	
	Swab		Swab		Swab	
	TP	%A	TP	%A	TP	%A
Aspergillus flavus	1	11.1	1	7.14	-	-
A.niger	1	11.1	-	-	-	-
A.fumigatus	2	22.2	2	14.2	-	-
A. sydowi	-	-	-	-	2	20
A. awamori	-	-	-	-	-	-
A. flavipes	-	-	-	-	-	-
A. sp.	1	11.1	-	-	-	-
Cladosporium cladosporoides	-	-	1	7.14	-	-
Rhizoctonia solani	-	-	-	-	-	-
Bipolaris sp.	-	-	1	7.14	-	-
Paecilomyces sp.	-	-	-	-	2	20
Verticillium alboatrum	-	-	-	-	1	10
Curvularia geniculata	-	-	2	14.2	-	-
C. subulata	1	11.1	-	-	1	10
Spiromyces sp.	-	-	-	-	1	10
Rhizopus sp.	1	11.1	-	-	-	-
Alternaria alternata	1	11.1	1	7.14	1	10
A.solani	-	-	-	-	-	-
Trichoderma koningi	1	11.1	-	-	1	10
Penicillium sp.	-	-	2	14.2	1	10
Penicillium chrysogenum	-	-	4	28.5	-	-
Total	9		14		10	
Shannon diversity index	0.448		0.555		0.506	

TP =Total population Swabbing affected area size = 4x4 cm^2
%A= Percentage abundance - = absent

Overall analysis of fungal spectrum from Kailash temple showed that maximum Shannon diversity index was in rain season followed by winter and summer season.

Analysis of Deteriorated Stone Samples

Small deteriorated stone pieces (sample 1 to sample 18) were collected from Kailash temple, Agra fort and European Cemetry (S1 to S8 sandstone pieces from Kailash temple, S9 to S12 deteriorated sandstone pieces from Agra fort and S13 to S18 deteriorated sandstone pieces from European Cemetry) and qualitatively analyzed by direct plating method on Czapek's dox agar medium (table 4). The aim was to calculate the relative frequency of fungal species and to analyze the some undetected fungal species. Analysis showed the presence of about 23 fungal species, with maximum relative frequency 12.5% for *Curvularia geniculata* and *Bipolaris sp*. The next fungi in this sequence were *Rhizictonia solani* (r.f. 8.33%) and *Gilmaniella sp*. (r.f. 7.29%). The minimum relative frequency (r.f. 1.04%) was observed in case of *Microsporum sp.* (plate 2 Figure b), *Penicillium purpurrescens, Mucor sp., Fusarium sp.* and *Staphylotrichum sp.*

Photomicrographs of *Curvularia subulata* & Monilia sp. (from Agra fort Agra), *Trichoderma koningi* & *Alternaria alternata* (from European Cemetry, Agra), *V. alboatrum* (from Kailash temple Agra) and *Microsporum sp.* (from deteriorated sandstone sample) have been shown in Plate 2 figure a to f respectively.

Caneva et al. (1991) described *Cladosporium herbarum, Aspergillus niger, Stachybotrys* spp. and *Alternaria* spp. as stain causing microorganisms forming black deposits on the stone surface. In the present investigation above said fungi had been isolated from different samples.

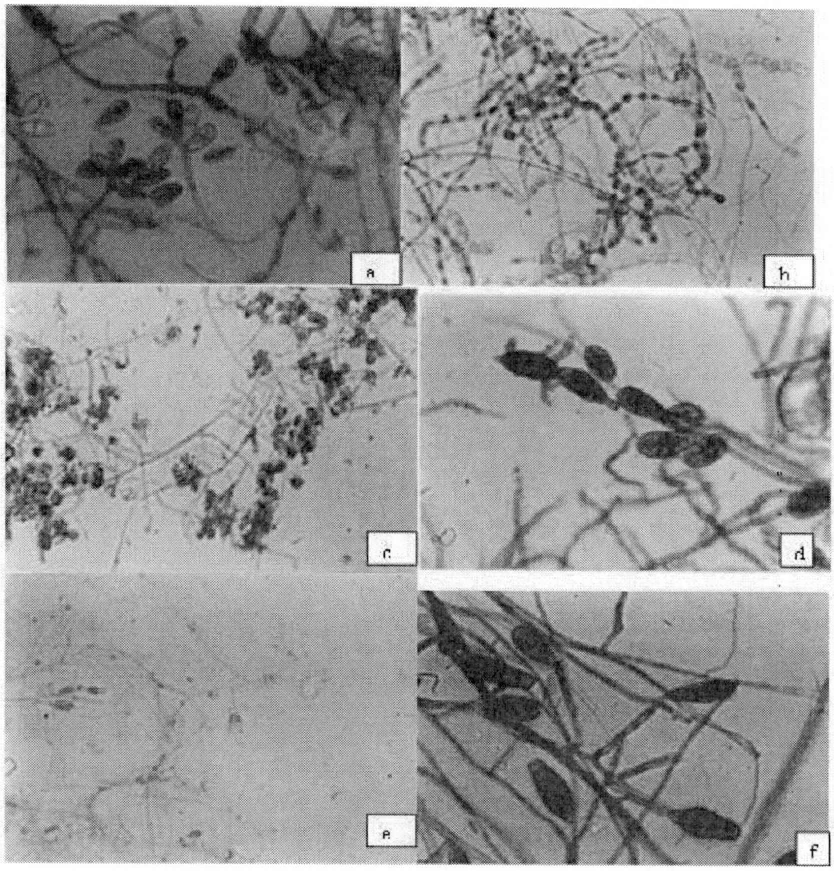

Plate 2. (a). Photomicrograph of *Curvularia subulata* from Agra fort
(b). Photomicrograph of *Monilia sp.* from Agra fort
(c). Photomicrograph of *Trichoderma koningi* from European Cemetry
(d). Photomicrograph of *Alternaria alternata* from European Cemetry
(e). Photomicrograph of *Verticillium alboatrum* from Kailash temple
(f). Photomicrograph of *Microsporum sp.* from deteriorated sandstone sample

Table 4. Relative frequency of fungi for deteriorated stone samples

Fungi	Relative Frequency (%)
Aspergillus flavus	5.20
A.niger	6.25
A.unguis	3.12
A.awamori	5.20
Penicillium livida	5.20
P. purpurrescens	1.04
Monilia sp.	5.20
Trichoderma koningi	4.16
Microsporum sp.	1.04
Curvularia geniculata	12.5
Bipolaris sp.	12.5
Rhizoctonia solani	8.33
Ambylosporium sp.	3.12
Streptomyces sp.	2.08
Mucor sp.	1.04
Rhizopus sp.	5.20
Fusarium roseum	2.08
Fusarium sp.	1.04
Gilmaniella sp.	7.29
Hyalodendron sp.	2.08
Staphylotrichum sp.	1.04
Papulospora sp.	3.12

Gomez-Alarcon (1995) was consistent with the experimental results and reported fungi belonging to the genera *Alternaria, Penicillium, Phoma, Trichoderma, Mucor, Ulocladium,* and *Dictyodesmium* colonizing the several monuments of granite, limestone & brick at Alcala de Henares (Madrid,Spain). Similarly Resende et al. (1996) reported that acid producing fungi include *Acremonium sp, Aspergillus puniceous, Fusarium oxysporum, Mucor hiemalis, Penicillium expansum* and *P. nigricans* which chemically deteriorate the stone surface.

In many observations it has been seen that smooth surfaces like freshly exposed polished marbles, were not easily colonized by microorganisms however after few years of exposition marble surfaces become rough and begin to be colonized by pigmented fungi like *Epicoccum purpurascens*, species of *Alternaria* and *Cladosporium*; and black yeasts like *Phaeococcus, Capnobotryella* etc. (Urzi et al.1999). Even, cleaning treatments with or without consolidation procedure could favor the fungal growth as primary colonizers like *Cladosporium, Alternaria* and *Nigrospora* (Wollenzien et al. 1995).

Many workers (Martinez et al. 2003; Herrera & Arroyave 2003; Martinez et al. 2003a) working on monuments of Colombia reported fungal genera *Cladosporium, Rhodotorula, Aureobasidium, Penicillium, Curvularia, Acremonium, Fusarium, Verticillium, Debaromyces, Stachybotrys, Gliocladium* and *Aspergillus niger*; *Rhizopus* sp. and *Mucor* sp.

A perusal of study and experiments had pointed out that extreme temperature is the only tolerated by prokaryotes. The temperature maximum known for eukaryotic organisms is between 50° and 60°C. This group is chiefly represented by fungi eg members of *Rhizopus, Penicillium, Humicola* and *Mucor*. Thermophilic fungi such as *Aspergillus fumigatus* and *Geotrichum candidum* are involved in the degradation up to temperature of 50-60°C. Then thermotrophic such as *Mucor pusillus* and *Humicola lanuginosa* come into action and further heating beyond 65°C is related to the activity of bacteria and actinomycetes, (Stolp 1988). It was also experienced that higher the temperature, the smaller the number of species present in extreme environment. In this sequence Lipson et al. (1999) reported that on average, microbial biomass was lower in the summer than at other times. Microbial biomass was highest when soils were cold, in autumn, winter and early spring.

Relative humidity is inversely proportional to temperature. Higher the temperature, lower the relative humidity. Therefore in the experimental city when the temperature goes as high as 40 – 45°C during summer, the relative humidity goes as low as 50-60%. During rainy season even the temperature is not much lower yet the relative humidity is highest and maintained due to rainfall. This is the reason when maximum diversity is found during this season. On the opposite end of summer, temperature is very low (0-25°C) and relative humidity is higher (70 – 85%) during the season of winter. This season also favours the growth of molds but lower temperature (<25 °C) creates obstacles in the path of diversity. One of the SEM photomicrograph is showing sporangiola of *Cunninghamella sp.* and cup shaped molds on stone surface (plate 3 Figure e).

Tropical Chamber Test

At first the sandstone pieces without the deposition of any organic matter or black patina or crust were washed with distilled water and put inside the tropical chamber at different relative humidity. After 6 months, the microscopic observations showed absence of any growth of mold. It was presumed that relative humidity cannot initiate the growth of mold unless organic compounds and quiescent fungal spores are present on the surface of stone.

In the next series of practical, the sandstone pieces with deposited black crust were kept in tropical chambers at 52%, 76%, 85% (plate 3 Figure a), 96.6% relative humidity. Within six months there was fibrous growth of molds on experimental surface of stone which belonged to the tropical chamber of R.H. 76%, 85% (plate 3 Figure b) & 96.6%. There was no fungal growth on the stone piece kept at 52% R.H. when observed from optical microscope. Molds were only qualitatively analyzed for their diversity (table 5). It was shown that diversity was maximum for those sandstone pieces kept at 96.6% R.H. followed by 85% R.H. During the study of 24 months 17 species appeared at 96% R.H., 14 species appeared at 85% R.H. and 6 species at76% R.H. (constant temperature of 25°C).

The species of molds which flourished at 76% R.H. are *Aspergillus flavus, A.terreus, A. fumigatus, A. sydowi, Alternaria alternata* and *Cladosporium cladosporoides* (table 5). The relative frequency was maximum for *A. sydowii* (36.3%) followed by *A. fumigatus* (27.2%), while rest of the four mold were having minimum relative frequency (9.09%).

At 85% relative humidity, *Fusarium moniliforme* and *Cephalosporium acremonium* were of 17.3% relative frequency (maximum value) followed by *Aspergillus flavipes* (r.f.= 13%). All the other remaining molds were of 4.34 to 8.69% relative frequency.

The relative humidity 96.6% is most favourable of the molds as reflected by number of molds. At this relative humidity *Aspergillus flavipes, Fusarium oxysporum* had 11.76% r.f. and *Aspergillus sydowi*, Cladosporium *cladosporoides* had 8.82%r.f., while rest of the 10 molds had r.f. in the range of 2.94% to 5.88%. Hence overall spectrum revealed that

Aspergillus sydowi and *A. fumigatus* proliferated at all the experimental relative humidity. The pure plate of *Verticillium puniceum* and photomicrograph of *Stachybotrys sp.* has been shown in plate 3 Figure c &d respectively.

The comparison between fungal diversity of monuments and tropical chamber showed that there is quite difference between the two. There is fluctuation of relative humidity and temperature on the surface of monument while in tropical chamber relative humidity and temperature are constant parameters. The perusal of study showed that species of *Aspergillus* like *A. flavus, A. niger* prefer 76% to 85% R.H. and on monuments they grow in summer and rain season. *A. fumigatus* and *A. sydowi* can be obtained in three seasons and at all the relative humidity. On general account *Penicillium sp., Fusarium sp., Paecilomyces sp., Stachybotrys sp., Verticillium sp., Cephalosporium acremonium* (exceptions are also present) required higher relative humidity on monument as well as in tropical chamber.

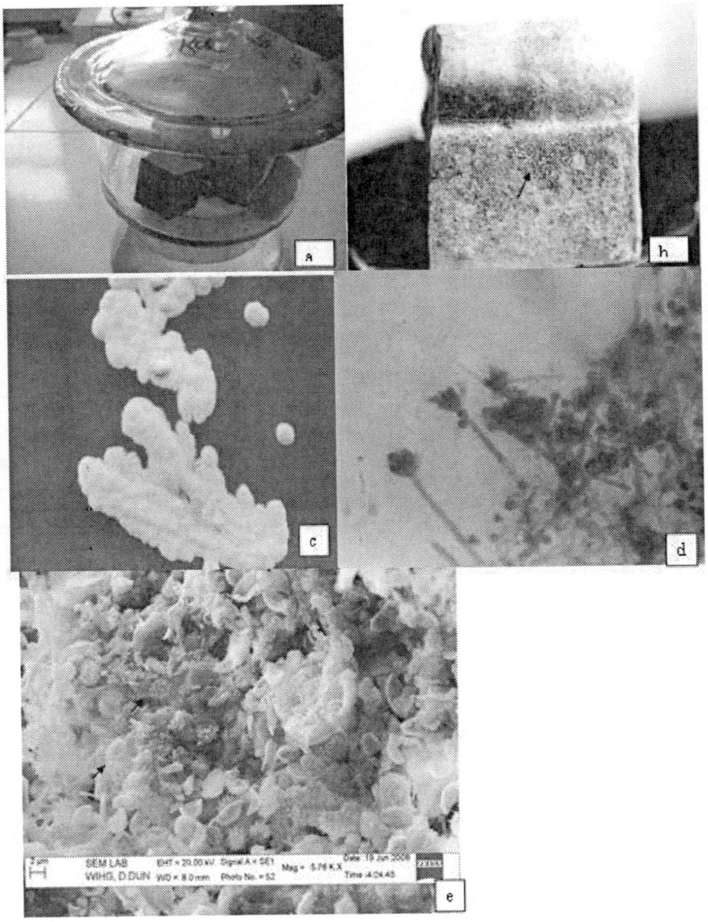

Plate 3. (a). Tropical chamber test – the sandstone pieces kept at 85% relative humidity
(b). The growth of fungi after 6 months on sandstone surface kept at 85% R.H.
(c). Pure plate of *Verticillium punicum* isolated at 96.6% relative humidity
(d). Photomicrograph of *Stachybotrys sp.* isolated at 96.6% relative humidity
(e). SEM Photomicrograph of naturally deteriorated sandstone obtained from Agra fort showing tiny sporangiola of *Cunninghamella sp.* and other cup shaped fruiting body of unknown mold (magnification x 5.76 KX)

Table 5. Relative frequency of fungi at different relative humidity (R.H.) level in tropical chamber test method

Fungi	Relative frequency (%)		
	76% R.H.	85% R.H.	96.6% R.H.
Aspergillus flavus	9.09	-	-
A. niger	-	4.34	-
A. terreus	9.09	-	-
A. fumigatus	27.2	4.34	2.94
A. flavipes	-	-	11.76
A. sydowi	36.3	13.0	8.82
A. unguis	-	4.34	2.94
A. awamorii	-	4.34	-
Alternaria alternata	9.09	4.34	2.94
Rhizoctonia solani	-	4.34	2.94
Cladosporium	9.09	4.34	8.82
cladosporoides	-	4.34	2.94
Stachybotrys sp.	-	-	2.94
Trichoderma	-	17.3	8.82
koningi	-	-	8.82
Fusarium	-	-	11.76
moniliforme	-	17.3	8.82
F. roseum	-	4.34	2.94
F. oxysporum	-	8.69	-
Cephalosporium	-	4.34	-
acremonium	-	-	2.94
Penicillium livida	-	-	5.88
Papulospora sp.	-	-	2.94
Mycelia sterilia			
Verticillium			
alboatrum			
V. puniceum			
Ganodobotrys sp.			
Total no. of species	6	14	17

Temperature and relative humidity are environmental factors important in determining whether fungi can subsist or grow on substrate. Most fungi may be grouped into three categories on a thermal basis viz. Psychrophiles, mesophiles and thermophiles. In general it has been shown that members of the Mucoracea and Deuteromycetes would grow on substrates stored at relative humidity of 90-100% and not below; however some species of *Penicillium* would grow at humidity as low as 75%. Members of the *Aspergillus glaucus* group have even been found growing in humidity as low as 65% R.H. (Bonner and Fergus 1960). In the present context it was noticed that members of deuteromycetes only appeared during the whole course of study and Penicillium livida proliferated not less than 85% RH.

Many of the scientists like Mishra (1981) reported that *Aspergillus flavus, A. fumigatus, A. niger, A. ochraceus, A. sydowi, Chaetomium dolichotrichum, C. indicum, C. globosum, C. biapiculatum, Penicillium oxalicum* and *Rhizopus arrhizus* were not affected at different relative humidity and temperature.It was also reported that for *Stachybotrys atra* the suitable relative humidity was 94%. These observations are all the same to that with experimental results of tropical chamber and molds of monuments. Some other authors Chauhan (2004), Bhawana (2004) working on deterioration of marble and paper material respectively reported the same results for the relative humidity.

Pattern of Deterioration

Molds and other microorganism make patina on the infected surface. Patina can be explained as microscopic visualization of complex structured communities – the biofilms which are constituted by microorganisms aggregated and sustained by a self synthesized extracellular matrix of polymeric substances. It differs in thickness, viability, density and organism composition. With reference to fungi, the biodeterioration can follow any or all of the following pathways:- aesthetic damage due to fungal pigments especially melanins and carotenoids (Urzi et al. 2000). biochemical diagenesis because of the dissolution of mineral substrates by organic acids (oxalate, citrate, malate, fumarate, propionate) that are exuded in the fungal growth environment. These acids also causes decohesion of calcite bound quartz grains in sandstone.

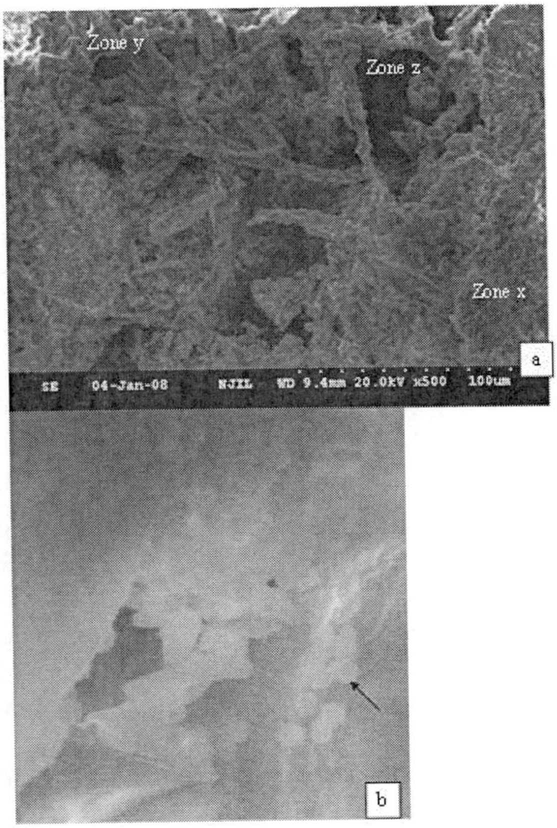

Plate 4. (a). SEM Photomicrograph showing three dissolution zone on sandstone surface (magnification x 500)
(b) SEM Photomicrograph showing biomineral embedded on the surface of fungal hyphae (magnification x 900: bar 50μm)

Intercrystalline growth and biopitting – The meristematic fungi (the cells of those fungi are thick walled and look like meristematic cells) specially involved in such type of procedure. The fungi grow following the weakest part along stone crystal causing their detachment and may form the pits. The SEM photomicrograph showed several dissolution zone (Plate 4 Figure a) that reflect different levels of dissolution. Some zone showed no dissolution (zone x), some showed medium degree of dissolution (zone y), and some showed severe dissolution (zone z). At some places hyphal growth is accompanied by new biomineral. These biomineral crystals are of different shapes and size. The crystals are either single, in pair or in cluster. These are engulfed within the fungal hyphae. These crystals were confirmed by XRD (Figure 1) and EDX (Figure 2). Energy dispersive X-ray analysis showed the peaks of silicon, iron, aluminium, chlorine, sodium, calcium and magnesium. There were approximately equal peaks of calcium and magnesium indifferent to that of original stone where calcium is more than magnesium. This result indicated toward the formation of biomineral.

X-ray diffraction revealed the highest peak of quartz, the main component of sandstone and small peaks of minerals (baratovite, feldspar, orthoclase, calcite) and organic pollutants (tetrachloroquinone, potassium formate, malonate, benzoate etc.). There was also a small dual peak of fungal precipitated dolomite similar to that of other reports (Kolo et al. 2007; Rodriguez-Navarro et al. 1997).

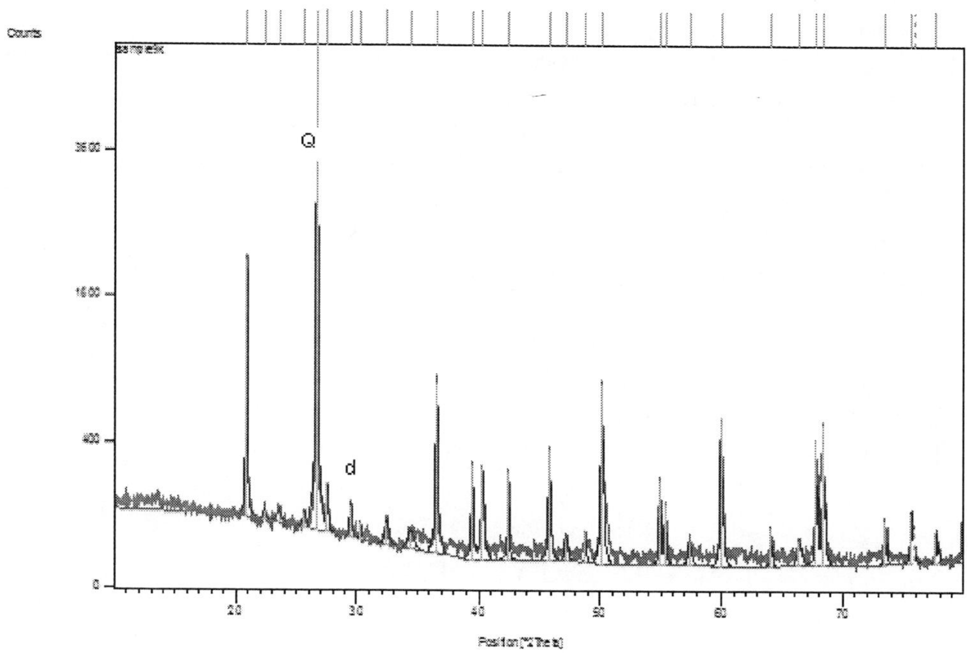

Figure 1. XRD of biologically weathered sandstone (Q = quart low; d=dolomite)

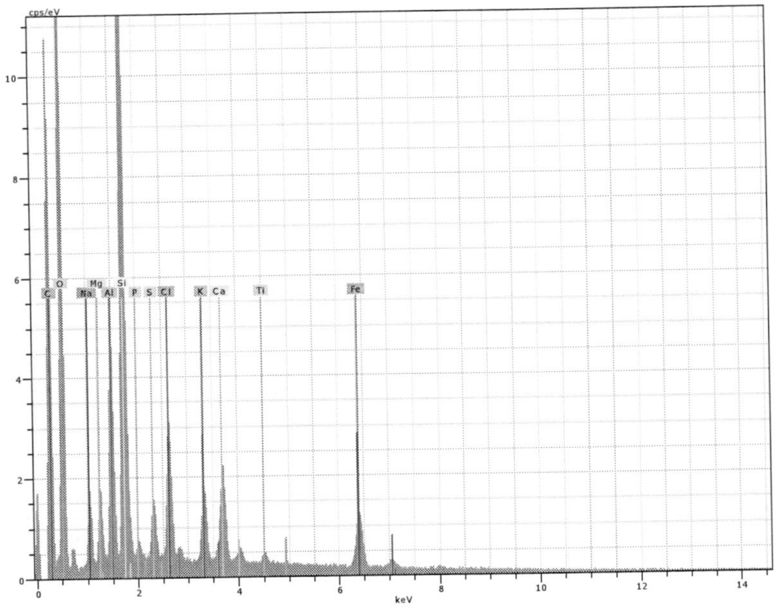

Figure 2. EDX of biologically weathered sandstone

There are many reports that fungi/molds are involved in biogenesis of crystals along with the mineral dissolution (Alexopoulos et al. 2004). The hyphae and in some cases the sporangiola of a variety of different types of fungi have been found to be decorated with crystals of calcium oxalate. Recently there is report of in-vitro formation of calcium oxalate crystals by ectomycorrhizal fungus *Piloderma fallax* (Tuason and Arocena 2009). In the case of certain litter inhabiting species aggregation of crystals may be as dense as to give mycelia a white, crust like appearance. When viewed with scanning electron microscopy fungal crystals often are spectacular in appearance. The precipitation of crystals of dolomite on the surface of fungal hyphae have also been observe and confirmed by Kolo et al. (2007), Rodriguez-Navarro et al. (1997) and Jain et al. (2009).

The organic acids exuded by fungi act as strong chelators and form complexes with various metals in the mineral substrate: calcium, magnesium, manganese, zinc, copper, aluminium and iron in a process of metal mobilization and immobilization that cause mineral dissolution and precipitation of new mineral (Gadd et al. 2002).

ACKNOWLEDGMENTS

Author thanks to Dr. V.M. Katoch (Director), Dr. Virendra Kumar National JALMA Institute of Leprosy and other Mycobacterial diseases (ICMR), Agra (India) for providing SEM facility; Dr. N.K.Saini, Dr. N.K.Juyel Wadia Institute of Himalayan Geology Dehradun (India) for XRD facility and SEM-EDX analysis. Special thanks go to the Principal R.B.S. College Agra and Dr. Seema Bhadauria Department of Botany, R.B.S. College Agra for providing research facility and technical assistance. This work is financially supported by Department of Biotechnology, New Delhi.

References

Alexopoulos, C. J., Mims, C. W. & Blackwell, M. (2004). *Introductory Mycology publisher*, John Wiley & sons. *869*.

Barnett, H. L. & Hunter, B. B. (1972). *Illustrated genera of imperfect fungi ed. III*. Burgess Publishing Company, *241*.

Beech, I. B. (2004). Corrosion of technical materials in the presence of biofilms-current understanding and state-of-the art methods of study. *International Biodeterioration and Biodegradation*, *53*, 177-183.

Bonner, R. D. & Fergus, C. L. (1960). The influence of temperature and Relative humidity on growth and survival silage fungi, *Mycologia*, *52(4)*, 642-647.

Caneva, G., Nugari, M. P. & Salvadori, O. (1991). *Biology in the conservation of works of Art*. Roma: ICC ROM.

Chauhan, A. (2004). *Microbial deterioration of Marble and Marble products*. Thesis Dr. B.R. Ambedkar University, Agra, *112*.

Domsch, K. H. Gams, W. & Anderson, T. H. (1980). *Compendium of soil fungi*. London: Academic Press.

Ellis, M. B. (1976). *More dematiaceous Hyphomycetes*, EMI, KEW, Surrey, England. *507*.

Gadd, G. M. (2002). Fungal influence on mineral dissolution and metal mobility: Mechanisms and biogeochemical relevance, paper presented at 12th *Annual Goldschmidt Conference*, Eur. Assoc of Geochem., Davos, Switzerland.

Gilman, J. C. (2001). A Manual of soil fungi. *Biotech Books*, Delhi. *392*.

Giullitte, O. (1995). Bioreceptivity: a new concept for building ecology studies. *The Science of the total environment*, *167*, 215-220.

Gomez-Alarcon, G., Cilleros, B., Flores, M. & Lorenzo, J. (1995). Microbial communities and alteration processes in monuments al Alcala de Henares, Spain. *The Science of total Environment*, *167*, 231-239.

Gorbushina, A. A., Heyrman, J., Dornieden, T., Gonzalez-Delvalle, M., Krumbein, W. E., Laiz, L., Peterson, K., Saiz-Jimenez, C. & Swings, J. (2004). Bacterial and fungal diversity and biodeterioration problems in mural painting environments of St. Martins Church. *International Biodeterioration and Biodegradation*, *53*, 13-24.

Herrera, L. K. & Arroyave, C. (2003). Atmospheric and biological deterioration of two churches from the cultural heritage of the city of Medellin, Colombia In. *Proceedings of the International Congress on Molecular Biology and cultural Heritage*, 4-7 March, 2003, Sevilla, Spain. (ed) C.Saiz-Jimenez. 271-276.

Jain, A., Bhadauria, S., Kumar, V. & Chauhan, R. S. (2009). Biodeterioration of sandstone under the influence of different humidity levels in laboratory conditions. *Building and Environment*, *44*, 1276-1284.

Kolo, K., Keppens, E., Preat, A. & Claeys, P. (2007). Experimental observations on fungal deagenesis of carbonate substrates. *Journal of Geophysical Research*, *112*, 1-20.

Kumari Bhawana, (2004). *Biodeterioration of library items and antique manuscript material*. Thesis, Agra: Dr. B.R. Ambedkar University.

Lipson, D. A., Schmidt, S. K. & Monson, R. K. (1999). Links between microbial population Dynamics and nitrogen availability in an alpine ecosystem. *Ecology*, *80(5)*, 1623-1631.

Martinez, M., Gaylarde, C. C. & Otalora, A. (2003). Preliminary microbiological analysis of biofilms of some monuments in Villa de Legva, Colombia. In C. Saiz-Jimenez, (ed). *Proceedings of the International Congress on Molecular Biology and Cultural Heritage*, 4-7 March, Sevilla, Spain 277-280.

Martinez, M., Martinez, P. C. & Laverde, P. (2003a). Microbiological studies of biofilm present on stones from the National Museum Building, Bogota, Colombia. In C. Saiz-Jimenez, (ed) P*roceedings of the International Congress on Molecular Biology and Cultural Heritage*, 4-7 March, Sevilla, Spain: 253-258.

Mishra, N. (1981). Influence of temperature and relative humidity on fungal flora of some species in storage. *Zeitschrift fur Lebensmitteluntersuchung and Forschung A, 172(1)*, 30-31.

Nigam, S. S. (1965). Laboratory test methods in microbiology, *issued by Defence Research Laboratory (Materials)*. Ministry of Defence, Kanpur.

Perry, R. S., Gorbushina, A., Engel, M. H., Kolb, Y. M., Krumbein, W. E. & Staley, J. T. (2003). Accumulation and deposition of inorganic and organic compounds by Micro-colonial fungi. *In-Proceeding of the third European workshop on exo-Astrobiology*, 18-20 Nov, 2003 Madrid Spain (ed.) R. A. Harris, & L. OaWeh, and ESA SP-545.

Resende, M. A., Resende, G. C., Viana, E. V., Becker, T. W. & Warsheid, T. (1996). Acid production by fungi isolated from historic monuments in the Brazilian State of Minas Gerais. In: *Biodegradation & Biodeterioration in Latin America.* 65-67. Porto Alegre: Mircen.

Rodriguez-Navarro, C., Sebastian, E. & Rodriguez-Gallego, M. (1997). An Urban model for dolomite precipitation: authigenic dolomite on weathered building stones. *Sedimentary Geology*, *109*, 1-11.

Roldan, M., Clavero, E. & Herandez-Marine, M. (2003). Aerophytic biofilms in dim habitats. In: Saiz-Jimenez, C. (ed.) Molecular Biology and Cultural heritage, *Proceedings of International Congress*, Sevilla. Balkema Publisher, Lisse (NL), 163-169.

Sterflinger, K. (2000). Fungi as geological agents *Geomicrobial J*, *17*, 97-124.

Stolp, H. (1988). *Microbial Ecology*: Organisms, habitats, Activities. (Publisher) Cambridge University Press, 324.

Tuason, M. M. S. & Arocena, J. M. (2009). Calcium Oxalate biomineralization by *Piloderma fallax* in response to various levels of calcium and phosphorus. *Applied and Environmental Microbiology*. Doi:10.1128/AEM. 00325-09.

Urzi, C., De Leo, F., Salamone, P. & Criseo, G. (1999). Impact of airborne fungi on marble objects exposed at Messina Euromarble Site. In snethlage (ed.) *Proceedings of 9th Euromarble Eurocare EU 496 Workshop*: 39-55. Munich.

Urzi, C., De Leo, F., de Hoog, & Sterflinger, K. (2000). Recent advances in the molecular biology and ecophysiology of meristematic stone-inhabiting fungi. In O., Ciferri, P. Tiano, & G. Mastromeri, (eds) of Microbes and Art: *The role of microbial communities in the degradation and Protection of Cultural Heritage*: 3-19 New York. Kluwer Academic/Plenum Publisher.

Warscheid, T. H. & Braams, J. (2000). Biodeterioration of stone: a review. *International Biodeterioration and Biodegradation*, *46(4)*, 343-368.

Wollenzein, U., De Hoog, G. S., Krumbein, W. E. & Urzi, C. (1995). On the isolation of micrcolonial fungi occurring on and in marble and other calcalreous rocks, *Science total Environment*, *167*, 287-294.

In: Relative Humidity: Sensors, Management...
Editor: John M. Castillo

ISBN: 978-1-61761-734-8

Chapter 7

RELATIONSHIPS BETWEEN TIME OF WETNESS, RELATIVE HUMIDITY, TEMPERATURE AND CORROSION OF METALS IN TROPICAL LOW-POLLUTED ATMOSPHERES

Juan G. Castaño, Juan Delgado and Felix Echeverría
Corrosion and Protection Group, University of Antioquia, Medellin, Colombia

ABSTRACT

In atmospheric corrosion the time of wetness (TOW) refers to the time when the metal surface is sufficiently wet for corrosion to occur. It depends on the climatic conditions and on the nature of the surface at hand, and usually is expressed as a function of relative humidity (RH) and temperature (T). According to ISO 9223 standard, atmospheres with concentration of SO_2 lower than 10 mg/m^{2}day and concentration of chlorides lower than 3 mg/m^{2}day are considered as low polluted environments, and the effect of these atmospheric agents in the corrosive attack is considered as insignificant. In this type of atmospheres, corrosion rate of metals is closely related with this parameter, which is generally high in tropical atmospheres. However, research works about the relationship between TOW and corrosion of metals in tropical atmospheres with low concentrations of SO_2 and chlorides are very scarce.

In this study, carbon and galvanised steel were exposed for 14 months (between 2006 and 2007) at 9 different tropical sites in Colombia. These sites have low contents of SO_2 and chlorides, and have RH values higher than 60%. RH and T were recorded continuously in all sites, and TOW was calculated from these two parameters. Finally, models for the corrosion rate (mass loss) of both metals as a function of TOW and exposure time were obtained. Also, the applicability and limitations of the ISO 9223 standard in these atmospheres are discussed.

Introduction

Atmospheric corrosion is an electrochemical process that occurs during alternating drying and wetting cycles, generally through thin electrolyte films connecting anodic and cathodic areas. The dominant anode reaction is the metal dissolution, while the dominant cathode reaction is the reduction of dissolved oxygen in the aqueous electrolyte. Atmospheric corrosion depends on three fundamental factors: the nature of the material, the characteristics of the surrounding atmosphere and the characteristics of the metal-air interface. In turn, the characteristics of the surrounding atmosphere, regarding the corrosivity, mainly depend on the meteorological parameters and contaminants as sulfur dioxide (SO_2) and chlorides (Cl) [1-3].

There are two constituents of the atmosphere that are essential to the process of corrosion: oxygen and water vapor, whose concentration is usually expressed in terms of relative humidity (RH).

The first stage of interaction between the metal and the atmosphere is the instant reaction of water vapor with the solid. The water molecule may bond in either molecular (through the oxygen atom to the metal or to another positively charged surface constituent) or dissociated form (through the formation of metal-oxygen or metal-hydroxyl bonds). Upon further exposure to the atmosphere, subsequent atmospheric water is adsorbed in the molecular form. The first layer of water has a high degree of ordering relative to the substrate. The second and third layers are more mobile with a higher degree of random orientation. Aqueous films ticker than three monolayers possess properties that are close to those of bulk layer [1].

The lower the RH, the lower the number of adsorbed monolayers of water on the metal surface [4]. At 20% RH, approximately one monolayer of water is adsorbed; at 40% RH, between 1.5 and 2 monolayers; at 60%, between 2 and 5 monolayers, and at 80%, between 5 and 10 monolayers [1]. The corrosion process is significant when RH exceeds a certain value, called "critical relative humidity" (RH_c) [5]. When RH_c is reached, about 3 monolayers of water are adsorbed on surface, which is sufficient for the electrochemical reactions take place.

Factors that promote the consolidation of moisture layer (such as roughness, the presence of particulate matter, hygroscopic substances or corrosion products) decrease RH_c, while wind and other factors affecting the evaporation of moisture may increase it [6-8]. For carbon steel, for example, it is reported RH_c values from 40% in presence of NaCl [9] to 80-85% in presence of SO_2 [10].

The wetting of metal surfaces is primarily due to the action of rain and dew. The time of surface wetness by dew is usually much greater than that of the rain [11]. However, the rain has been the subject of more studies, since it has some special features. Their action on the metal surface can have an accelerating or retarding effect on the corrosion process. In the first case, because it carries pollutants active species to the metal surface, wets the metal surface, increases the time of wetness and facilitates the adsorption of pollutants. In the second case, because it washes the metal surface and carries contaminants and corrosion products [12].

Temperature (T) has a dual effect on atmospheric corrosion, as it can accelerate the process, and also can retard or even inhibit it [13], so is very hard to try to establish correlations between T and corrosion rate.

The most important effect of T is perhaps their influence on the formation of the layer of moisture on the surface. For a given concentration of water in the air, a decrease in T can consequently cause an increase in RH, which in some cases implies that the RH_c is reached and the corrosion starts. On the other hand, the increase in T can produce lower moisture condensation on the surface, or affect the stability and permanence of the moisture film [13, 14].

Although the concepts of RH and T are essential for understanding the phenomenon of atmospheric corrosion, the time of wetness (TOW) is a more accurate indication of the influence of humidity on the corrosion of metals.

TOW refers to the time or fraction of total exposure time during which the metal is wet enough for corrosion to occur. It depends on climatic factors such as T and RH, rain duration and intensity, sun radiation, direction and intensity of wind and the presence of hygroscopic particles such as chlorides [1, 15, 16].

The possible range for the values of TOW is from 0 to 8766 hours per year or from 0 to 1 when measured in time fraction.

According to ISO 9223 standard [17], TOW is a climate parameter, independent of the surface on which water is adsorbed, and is defined as a function of time, T and RH. In this standard, TOW is estimated as periods of time with RH higher than 80% and T above 0°C, simultaneously. Such a calculated TOW value does not necessarily correspond to the actual time of wetness, because the latter is determined by the effect of many factors [18]. In tropical climates it is common to have a combination of high T, high humidity and considerable precipitation, so the TOW is often very high [19-23].

The 'true' TOW is complicated to define and calculate. However, there are relatively single quantitative relationships for calculation of TOW as a function of annual averages of T and RH, which can be used for classification purposes [24]. The quantity is easy to calculate from measured climatic parameters and have a high correlation to the true TOW. A procedure that needs daily or more frequent values of both T and RH seems unnecessarily complicated and it should be preferable if annual averages of the individual parameters could be used in the calculation of TOW. For the other hand, the use of electronic sensors that indicate whether the surface is wet or not can allow greater accuracy in this determination [25].

The atmospheric corrosion of carbon steel has been subject of intensive research, and is influenced principally by TOW, SO_2 and Cl [15, 16, 26, 27]. In low polluted atmospheres, the corrosion rate increases with TOW and RH [28]. Generally, the corrosion rate is between 4 and 65 μm/year in rural atmospheres, 23 to 71 μm/year in urban environments, 26 to 175 μm/year in industrial sites and 26 to 104 μm/year in marine atmospheres [1].

The rust layer formed on carbon steels has a complex morphology. It is generally porous, with poor adherence, and cracked in its outer part. This structure gives a poor protection to steel against atmospheric corrosion, because it is not a barrier to the passage of electrolyte and corrosive species towards the metallic substrate [29, 30]. The rust is mainly composed of lepidocrocite (γ-FeOOH), which is usually the first oxyhydroxide formed, and goethite (α-FeOOH) [31-33]. Often, there is also maghemite (γ-Fe_2O_3) [34]. In marine atmospheres is common to find magnetite (Fe_3O_4) and akaganeite (β-FeOOH) [33-35].

In galvanised steels, zinc coating exhibits better atmospheric corrosion resistance than steel substrate. Steel is protected because zinc coating prevents the direct contact with aggressive species of the medium and, when there are discontinuities, pores or cracks in the coating, zinc exerts a galvanic or cathodic protection in these sites, forming a dense and adherent film of corrosion products that fills discontinuities [36, 37].

The crucial SO_2 effect in zinc atmospheric corrosion has been corroborated through a large number of studies [37-40]. When the RH_c is reached, SO_2 adsorbed in the humidity layer that forms on zinc is transformed into sulfite and, subsequently, into sulfate. The humidity layer is acidified by the oxidation of SO_2 to sulfate, which greatly accelerates the rate of corrosion. Generally, the corrosion rate is between 0.2 and 3 μm/year in rural atmospheres, 2 to 16 μm/year in urban and industrial environments and 0.5 to 8 μm/year in marine atmospheres. These rates are rapid enough that the original zinc surface is covered with corrosion products in hours or days [1].

Basic zinc salts as hydroxychloride, hydroxysulfate and sodium zinc chlorohydroxysulfate are commonly formed in polluted environments, depending on the SO_2 and Cl content [41-43]. Other compound found in zinc and galvanised steels are zincite (ZnO), zinc hydroxide ($Zn(OH)_2$) and hydrozincite ($Zn_5(CO_3)_2(OH)_6$) [39, 41].

ISO 9223 specifies five categories (C1 to C5) of the atmospheric corrosivity toward carbon steel, zinc, and other metals; where C5 is the highest category. In the last 15 years, different researchers have found still higher corrosion rates that actually fall beyond the classification upper limits of the standard [18, 19, 22]. There are two ways to assess the corrosivity of a given atmosphere: by measuring the corrosion rates of standard samples or by measuring the environmental parameters and assessing the corrosivity from tables.

The environmental parameters used in assessing the corrosivity are: TOW, SO_2 and air salinity (Cl). TOW is divided into the five categories showed in Table 1 [17].

SO_2 and Cl deposition rates are divided into four categories (Table 2 and 3). For the purposes of the standard, the deposition rates are expressed as annual average. Any concentration of SO_2 within category P_0 or Cl within category S_0 is considered in the standard as the background concentration and, hence, insignificant from the point of view of corrosive attack.

Table 1. TOW categories according to ISO 9223 [17]

Category	h/year	Time fraction
τ_1	≤10	≤0,001
τ_2	10-250	0,001-0,03
τ_3	250-2500	0,03-0,3
τ_4	2500-5500	0,3-0,6
τ_5	>5500	>0,6

Table 2. SO_2 categories according to ISO 9223 [17]

Category	Deposition rate of SO_2 (mg/m^{2}day)
P_0	≤10
P_1	10-35
P_2	35-80
P_3	80-200

Table 3. Cl categories according to ISO 9223 [17]

Category	Deposition rate of Cl (mg/m^2day)
S_0	≤3
S_1	3-60
S_2	60-300
S_3	300-1500

Experimental results have been demonstrated the limitations of this standard in many environmental conditions [18, 19, 21, 44, 45], and there have been different proposals for its modification [45-47]. However, in low polluted atmospheres the calculation of TOW following the standard proves usually to be sufficient [18, 19], although research works about the relationship between TOW and corrosion of metals in tropical atmospheres with low concentrations of SO_2 and chlorides are very scarce.

Colombia is located near the equator. Its territory is crossed by the Andes mountain range and has a wide geographical diversity, ranging from tropical and jungle regions to moorlands and snowy peaks. Until now, there are no studies to establish the applicability and limitations of the standard and the influence of TOW on the atmospheric corrosion of metals in low polluted Colombian atmospheres.

In this study, carbon and galvanised steel were exposed for 14 months at 9 different tropical sites in Colombia. These sites have low contents of SO_2 and chlorides (Categories P_0 and S_0, respectively, according to ISO 9223), and most of them have high TOW values. RH and T were recorded continuously in all sites and TOW was established for each site with these data. Finally, models for the corrosion rate (mass loss) of both metals as a function of TOW and exposure time were obtained using multivariate analysis methods (principal components and regression analysis).

FIELD EXPOSURE

Plates (with the dimensions of 100 mm x 150 mm x 2 mm) of commercial AISI-SAE 1016 carbon steel and hot-dip galvanised steel (with average coating thickness of 60 μm) were used for the field tests. Surface preparation of the plates consisted of degreasing with acetone and washing with detergent, cleaning with ethanol during 15 min and drying in a hot air jet. Following their preparation, the specimens were stored for 24 h in a desiccator. Immediately before testing, they were weighed on an analytical balance of 0.01 g of sensitivity.

Plates were exposed in the 9 sites shown in Figure 1, at an angle of 45° from horizontal and with skyward surface facing south, according to ASTM G-50 [48]. The study was carried out over a 14-month period between September 2006 and November 2007. Test specimens were retrieved each 2 months. During every sampling period, three samples were picked up from each site for weight loss test.

Figure 1. Location of Colombia and the atmospheric corrosion stations

The deposition rate of SO_2 in each station was determined using the lead dioxide sulfation candle, according to ASTM D-2010 standard [49]. The deposition rate of chlorides was determined using the wet candle method described in ASTM G-140 [50]. The analyses of chlorides and SO_2 were carried out bimonthly over the 14-month period of exposure. In addition, T and RH were recorded hourly for determining TOW in each site. TOW was estimated as periods of time with RH higher than 80% and T above 0°C, simultaneously.

For weight loss measurements, corrosion products on the specimen surfaces were removed chemically by immersion in a specific solution, according to ASTM G-1 [51]. After corrosion products had been completely removed, the specimens were rinsed with distilled water, dried with blower, and then weighed to determine their mass loss.

The statistical analysis for the data obtained in the field tests was performed using multivariate analysis methods (principal components and regression analysis).

Environmental Data at Exposure Sites

Annual mean environmental data and ISO 9223 classification for the nine sites is shown in Table 4. Annual mean RH in almost all stations is high and the sites with the highest RH have at the same time the highest TOW. Several stations show a combination of high RH and high T. TOW is very high in two sites: Chinu (which is a site with very high RH, very high T, and has the highest deposition rate of chlorides) and La Union (which is a site with very high RH and relatively low T). The other stations have high TOW, with the exception of Pasto and Ocaña, where TOW is in τ_3 category.

Figure 2 shows the chloride deposition rate for the first level established in ISO 9223 (S_0) vs. the SO_2 deposition rate for the first level established in the same standard (P_0). It also shows the nine exposure sites in accordance with the contents of both pollutants. Regarding SO_2, most sites are located in the lower third of the category P_0, except for sites 1 (Barrancabermeja) and 9 (Ocaña), located in the middle third. For the chlorides, site 7 (Chinu) is located in the upper third of the category S_0 and site 8 (La Reforma) in the middle third. The rest are in the lower third.

Figure 3 show the average RH obtained in the nine sites for the seven bimonthly periods of evaluation. There is a great variability in this parameter, with variations of up to 20% from one period to another for sites like Barrancabermeja, La Reforma or La Virginia. There are a few things to be highlighted, such as the peak presented in the average RH value (95%) of site 1 (Barrancabermeja) that corresponded to the monitoring of the second period of the evaluation. Another one is that in all evaluation periods RH is higher than 60%.

Figure 4 shows the average T at the nine sites for various exposure periods. Temperatures did not exhibit major variations during the year in all sites. The little variation in the average temperature indicates that seasonal changes have a low incidence on it. In Colombia, these changes are distinguished primarily by the presence or absence of rain.

Table 4. Annual mean environmental data and ISO 9223 classification

#	Site	RH, %	T, °C	TOW	SO_2, mg/m^2day	Cl, mg/m^2day	ISO category
1	Barrancabermeja	83.1	28.0	0.576	3.84	0.18	$S_0P_0\ \tau_4$
2	San Felipe	75.0	29.2	0.500	2.46	0.74	$S_0P_0\ \tau_4$
3	La Virginia	68.7	25.7	0.333	0.67	0.36	$S_0P_0\ \tau_4$
4	San Bernardino	74.7	20.2	0.490	1.73	0.29	$S_0P_0\ \tau_4$
5	La Union	82.1	15.3	0.636	0.68	0.65	$S_0P_0\ \tau_5$
6	Pasto	70.1	15.1	0.255	0.60	0.50	$S_0P_0\ \tau_3$
7	Chinu	85.9	28.5	0.748	0.78	2.38	$S_0P_0\ \tau_5$
8	La Reforma	79.4	23.0	0.574	1.69	1.24	$S_0P_0\ \tau_4$
9	Ocaña	70.2	24.1	0.256	5.48	0.57	$S_0P_0\ \tau_3$

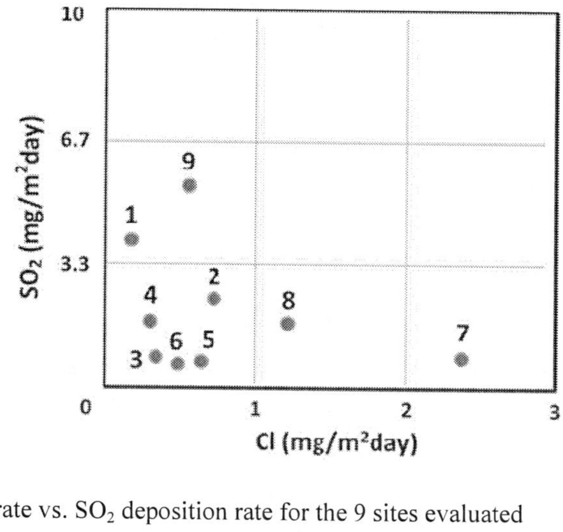

Figure 2. Cl deposition rate vs. SO_2 deposition rate for the 9 sites evaluated

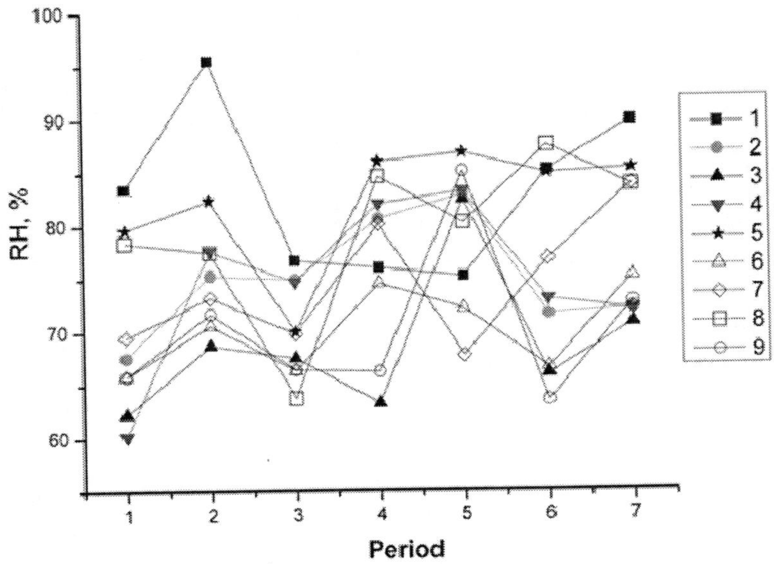

Figure 3. Average RH obtained in the 9 sites

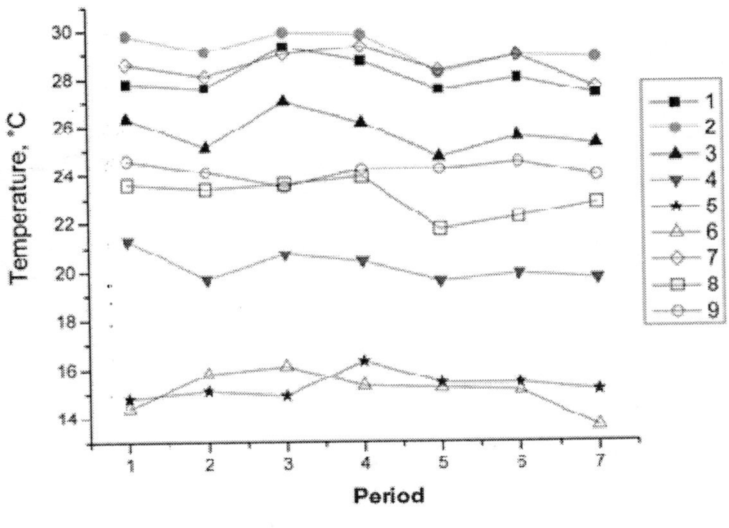

Figure 4. Average T at the 9 sites

Figures 5(a) and 5(b) show the TOW measured as fraction for each period evaluated. In all sites an increase of TOW around periods 4 and 5 is observed, which is consistent with the winter season recorded in the country in early 2007. In site 9 (Chinu) the highest TOW (0.9) was measured in period 5 (Figure 5(a)). This parameter reflects more than RH the seasonal changes occurred during the exposure period. In sites 3 (La Virginia), 4 (San Bernardino), 6 (Pasto) and 9 (Ocaña) very low TOW values were registered in the first exposure period (Figure 5(b)).

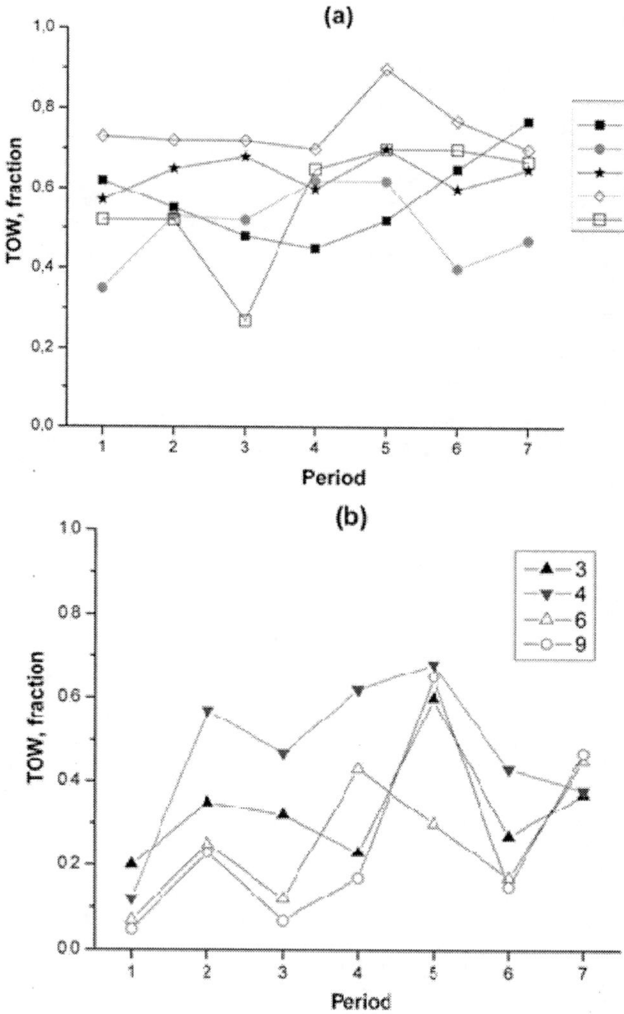

Figure 5. Variation of TOW measured as fraction for each evaluation period in: (a) sites with higher TOW; (b) sites with lower TOW

Corrosion Rates

In Table 5, annual corrosion rates and ISO 9223 classification of corrosivities for carbon and galvanised steel are presented. For carbon steel, Barrancabermeja is the most aggressive environment (C3 category), while other sites are in C2 category. However, the corrosiveness in Chinu approaches the threshold of C3 category.

For galvanised steel, most sites fall into the C3 category, except for Pasto and Ocaña.

Figure 6 shows the increase of mass loss of carbon steel in the 7 evaluation periods. From the first period, the stations of Barrancabermeja (1) and Chinu (7) show more corrosion than the others. At the end of the study, the mass loss in these sites is approximately three times higher than in the other. It is possible that this marked difference is influenced by factors other than the climatic parameters, which could be the contents of chlorides or SO_2.

Table 5. Annual corrosion rate and ISO 9223 classification at exposure sites

#	Site	Corrosion of carbon steel (μm/year)	Classification ISO 9223	Corrosion of galvanised steel (μm/year)	Classification ISO 9223
1	Barrancabermeja	25,78	C3	1,32	C3
2	San Felipe	4,23	C2	0,86	C3
3	La Virginia	5,64	C2	0,76	C3
4	San Bernardino	6,4	C2	0,93	C3
5	La Unión	4,67	C2	0,83	C3
6	Pasto	3,12	C2	0,57	C2
7	Chinu	22,92	C2	0,82	C3
8	La Reforma	5,29	C2	1,5	C3
9	Ocaña	7,97	C2	0,66	C2

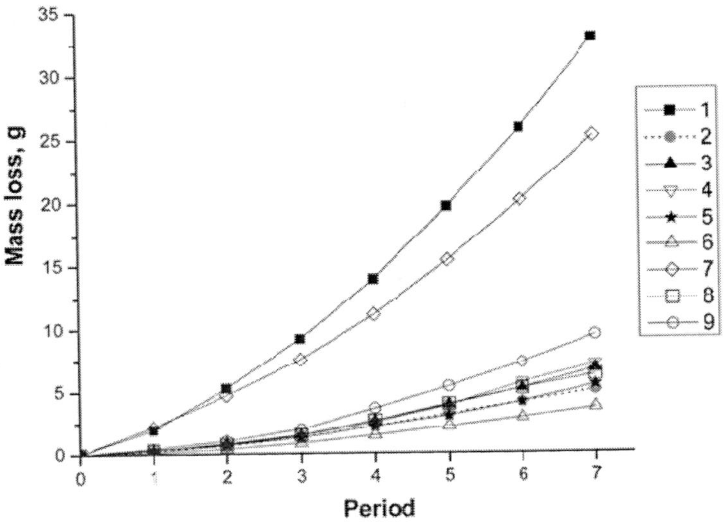

Figure 6. Mass loss of carbon steel at the 9 sites

In sites 1 (Barrancabermeja) and 7 (Ocaña), SO_2 annual mean deposition rates of 3.84 and 5.48 5.48 mg/m^{2}day were obtained. It seems that for this content of SO_2, there is a slight effect of this pollutant on the rate of corrosion, although these values are included in the lowest category of ISO 9223. Although in Ocaña the content of SO_2 is higher than in Barrancabermeja, it is possible that the corrosion rate in the latter site is greatly enhanced by the large difference in TOW from both stations (in Barrancabermeja is more than 2 times greater).

Site 2 (Chinu) is located to approximately 40 km from the coast, and the annual mean deposition rate of Cl was 2.38 mg/m^{2}day. Although this value is in the category S_0 of ISO 9223, is very close to the threshold of category S_1 (3 mg/m^2.d). In this case it is also probable that the Cl content is sufficient to significantly influence the corrosion rate recorded at this site for steel, which is the second highest of all sites considered.

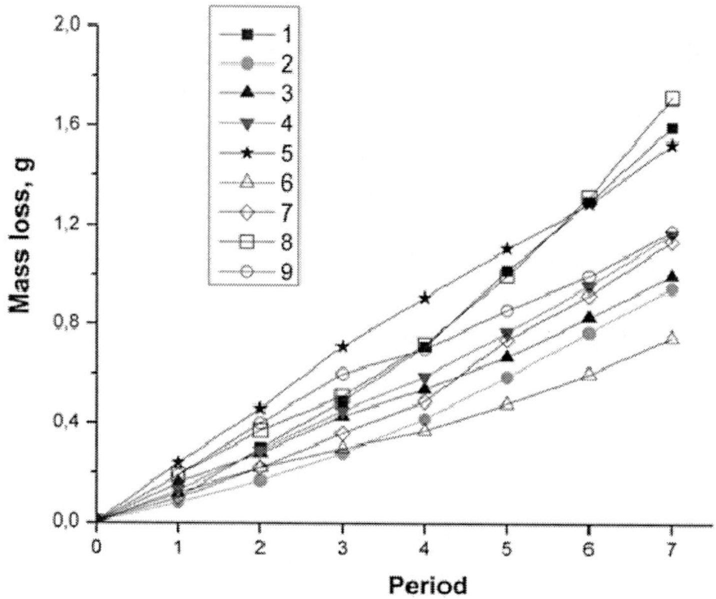

Figure 7. Mass loss of galvanised steel at the 9 sites

The observations in Barrancabermeja and Chinu indicates that, contrary to what is stated in ISO 9223, not always the concentration of SO_2 within category P_0 or Cl within category S_0 can be considered insignificant from the point of view of corrosive attack on carbon steel.

Figure 7 shows the increase in mass loss of galvanised steel in the seven periods evaluated. In this case, toward the end of the test a greater corrosion in sites 1 (Barrancabermeja), 5 (La Union) and 8 (La Reforma) is observed, although the differences from other sites are not very significant. Unlike carbon steel, in this case there is no influence of slightly higher contents of pollutants. Apparently thre is a greater influence of TOW, as in stations 1, 5 and 8, the TOW is slightly higher towards the end of the study.

Statistical Analysis for Carbon Steel

Due to the high variability in RH data (Figure 3) and the little variation in average T at each station (Figure 4), it was decided to develop models with variables that may accumulate over time. With accumulated data, it is possible to achieve greater accuracy and high correlation coefficients [52]. For these reasons, accumulated TOW was included as the only environmental variable. Therefore, for the statistical analysis, bimonthly data of exposure time (days), accumulated TOW (annual fraction) and mass loss of carbon steel (g) were used.

First, an analysis by principal components was done to obtain new variables or functions that are a linear combination of the TOW and exposure time variables, and explain most of the variability of the data with minimal loss of information, decreasing the dimension of the data. In this case, a single new variable has been obtained, named N. It accounts for 89.85% of the variability in the original data and its equation is:

$$N = 0.707*TOW + 0.707*t \qquad (1)$$

Where the values of TOW and t (exposure time) have been standardized by subtracting their means and dividing by the standard deviation. The two original variables have a similar weight in the new variable N.

Figure 8 illustrates the relationship between the new variable N and the square root of mass loss of carbon steel in the 9 sites. The square root of the dependent variable is used to correctly specify the model in its functional form.

The comparison of the regression lines shows that sites 1(Barrancabermeja), 7 (Chinu) and 9 (Ocaña) are statistically different from the others regarding to slope and intercept. For the 6 remaining sites (2, 3, 4, 5, 6 and 8), there are not statistically significant differences between the intercepts and/or slopes (Table 6 and 7). Therefore, it is possible to formulate a model of carbon steel corrosion for the six stations of similar behavior that suggests a group. For stations 9, 7 and 1 would have to formulate their respective models.

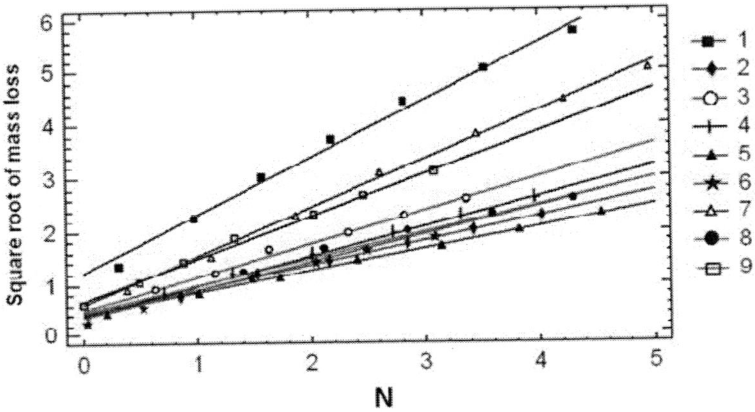

Figure 8. Regression lines for N vs. square root of carbon steel mass loss

Table 6. Coefficients of linear regressions

Site	Intercept	Slope	P-Value
2	2.22294	1.23715	0.0000
3	3.55599	1.90702	0.0000
4	2.89487	1.63014	0.0006
5	1.893	1.14488	0.0759
6	2.23684	1.15229	0.9411
8	2.6325	1.53	0.0271

Table 7. Analysis of variance for linear regressions (sites 2, 3, 4, 5, 6 and 8)

Source	Sum of Squares	Degrees of freedom	Mean Square	F-Ratio	P-Value
Model	158.817	11	14.4379	135.02	0.0000
Residual	3.20784	30	0.106928		
Total (Corr.)	162.025	41			

The mass loss of carbon steel in Barrancabermeja is the highest of all sites. For the other hand, the mass loss in Ocaña is slightly higher than in the others, except Chinu and Barrancabermeja, which is enough reason to not be grouped with the others. The reasons for the different behavior of these three stations were discussed in the previous section.

Finally, equation (2) corresponds to the model found for the corrosion of carbon steel in stations 2, 3, 4, 5, 6 and 8:

$$Mcs^{0.5} = 0.451 + 0.486N + (AR(1)=0.792) \tag{2}$$

Where Mcs is the mass loss of carbon steel. The adjusted model must have had a power transformation for Mcs of 0.5 for a correct specification of the functional form of the model. Furthermore, because it is data over time, the model also showed the presence of autocorrelation, which creates problems in the estimates of the model in the sense of losing their properties of unbiasedness and minimum variance. Therefore, it should be corrected to obtain the best unbiased estimators by the Cochrane-Orcutt method, including the AR(1) expression [53].

Table 8 shows some statistical estimates that prove the validity of the model.

The correlation coefficient (R^2) for this model is 96.86%, while the Durbin-Watson statistic is 1.73 (P=0.0000). The latter criterion tests the residuals to determine if there is any significant correlation based on the order in which they occur in your data file. Since the P-value is less than 0.05, there is an indication of possible serial correlation at the 95.0% confidence level.

Figure 9(a) show the relationship between N and mass loss of carbon steel, including data obtained experimentally and the fitting curve for the model. In Figure 9(b) the residuals are plotted with respect to N. No pattern is observed in their distribution, which serves to verify that the model is well formulated. However, constant variance but must be validated.

Table 8. Statistical estimates for the carbon steel model (equation (2))

	Coefficient	Std. Error	t-Statistic	Prob.
N	0.486220	0.014025	34.66869	0.0000
constant	0.451507	0.096332	4.687003	0.0000
AR(1)	0.791898	0.099803	7.934608	0.0000

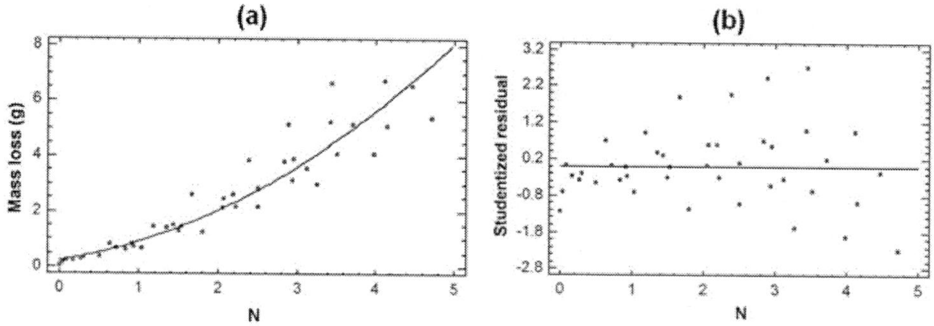

Figure 9. (a) N vs. mass loss and plot for the carbon steel corrosion model; (b) Residual plot

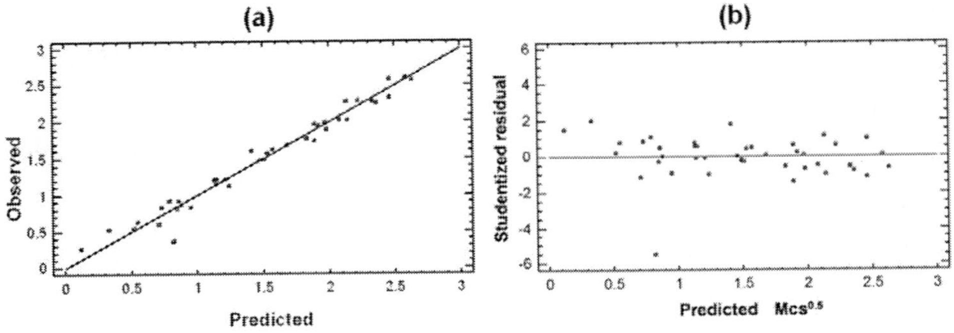

Figure 10. (a) Predicted data vs. observed data for carbon steel corrosion; (b) Residual plot for data predicted by the model

In Figure 10(a), data predicted by the model and experimentally observed are compared. A relatively low dispersion is observed, taking into account the variability of data measured in the field. For the other hand, in Figure 10(b), residuals are plotted with respect to the values predicted by the model transformed to the power 0.5. It is observed that the dispersion does not follow a regular pattern.

Statistical Analysis for Galvanised Steel

For the statistical analysis, bimonthly data of exposure time (days), accumulated TOW (annual fraction) and mass loss of galvanised steel (g) and were used. The variable N (equation (1)) is valid also for the analysis of the corrosion of galvanised steel

Figure 11 shows the relationship between the new variable N and the square root of mass loss of galvanised steel in the 9 sites. There are not statistically significant differences between the intercepts and/or slopes (Table 9 and 10). Therefore, it is possible to formulate a model of galvanised steel corrosion, statistically valid for all sites.

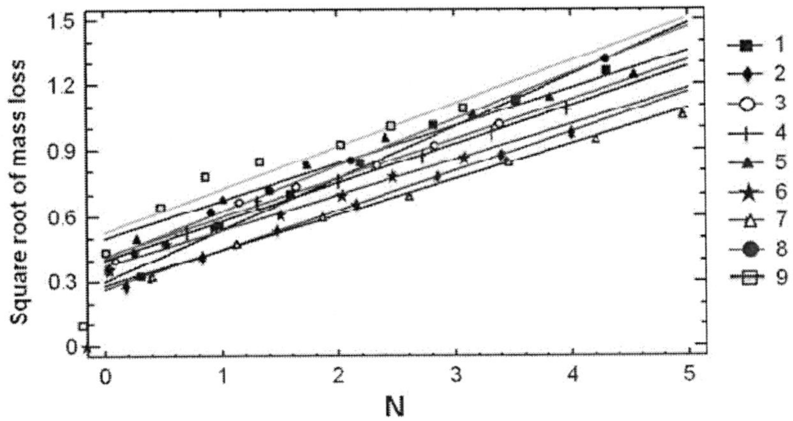

Figure 11. Regression lines for N vs. square root of mass loss of galvanised steel

Table 9. Coefficients of linear regressions showed in Figure 13

Site	Intercept	Slope	P-Value
1	0.708134	0.384552	0.0000
2	0.44734	0.227523	0.0000
3	0.638957	0.251043	0.0061
4	0.629163	0.259153	0.0016
5	0.778737	0.299662	0.0051
6	0.512817	0.199664	0.0000
7	0.424482	0.227942	0.0000
8	0.773626	0.371635	0.0081
9	0.881114	0.307503	0.0000

Table 10. Analysis of variance for linear regressions showed in Figure 13

Source	Sum of Squares	Degrees of freedom	Mean Square	F-Ratio	P-Value
Model	10.1549	17	0.597345	311.73	0.0000
Residual	0.08623	45	0.001916		
Total (Corr.)	10.2411	62			

The model found for the corrosion of galvanised steel was the following:

$$Mg^{0.5} = 0.417 + 0.183N + (AR(1)=0.84) \tag{3}$$

Where Mg is the mass loss of galvanised steel. As with carbon steel, the adjusted model must have had a power transformation for Mg of 0.5. This model also showed the presence of autocorrelation, which creates problems in the estimates of the model, as already mentioned in the previous chapter. Therefore, it should be corrected by the Cochrane-Orcutt method, including the AR(1) expression.

Table 11 shows some statistical estimates that prove the validity of the model.

The correlation coefficient (R^2) for this model is 95.04%, while the Durbin-Watson statistic is 1.60 (P=0.0000).

Figure 12(a) show the relationship between N and mass loss of galvanised steel, including data obtained experimentally and the fitting curve for the model. In Figure 12(b), the residuals are plotted with respect to N. No pattern is observed in their distribution, which serves to verify that the model is well formulated.

Table 11. Statistical estimates for the galvanised steel model (equation (3))

	Coefficient	Std. Error	t-Statistic	Prob.
N	0.183124	0.005165	35.45457	0.0000
constant	0.416865	0.051260	8.132346	0.0000
AR(1)	0.846569	0.069695	12.14668	0.0000

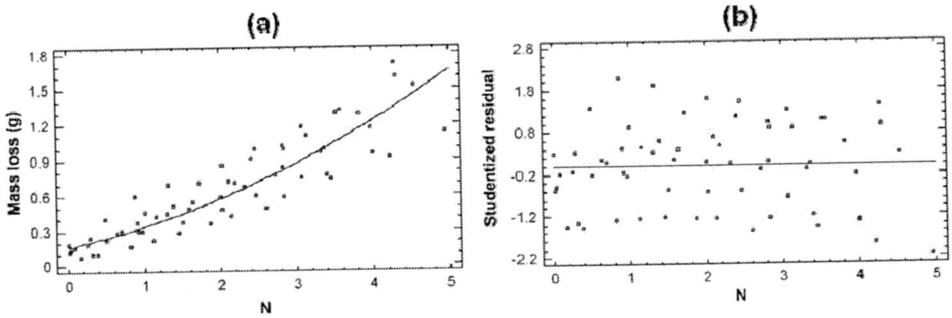

Figure 12. (a) N vs. mass loss for galvanised steel and plot for the model; (b) Residual plot for variable N

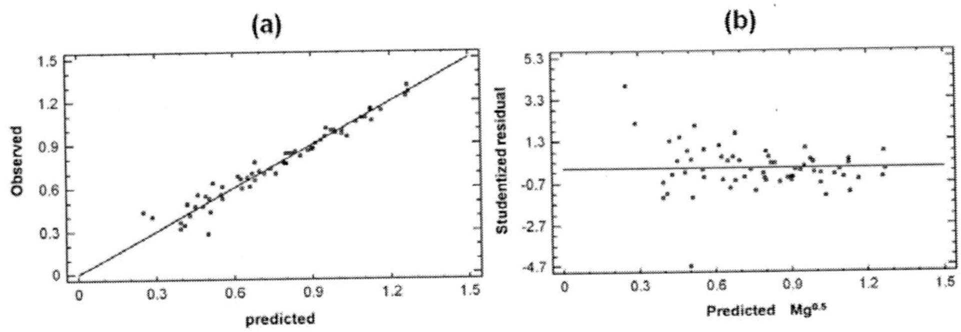

Figure 13. (a) Predicted data vs. Observed data for galvanised steel corrosion; (b) Residual plot for data predicted by the model

In Figure 13(a), data predicted by the model and experimentally observed are compared. A relatively low dispersion is observed, taking into account the variability of data measured in the field. Residuals are plotted in Figure 13(b) with respect to the values predicted by the model transformed to the power 0.5. It is observed that the dispersion does not follow a regular pattern, although the presence of atypical values should be assessed.

CONCLUSION

In tropical sites with low deposition rates of SO_2 and Cl (Category S_0P_0, according to ISO 9223 standard) it is possible to obtain models for the atmospheric corrosion of carbon and galvanised steel as a function of TOW and exposure time, with very high correlation coefficients. In the sites evaluated, bimonthly average RH showed high variation and TOW reflects more than RH the seasonal changes occurred during the exposure period. For carbon steel, the model obtained was not applicable at two sites with SO_2 deposition rates corresponding to the middle third of category S_0P_0. On the other hand, the same model was not applicable at sites with Cl deposition rates near the threshold of 3 mg/m^2d, which separates the categories S_0 and S_1. These observations indicate that, contrary to what is stated in ISO 9223, not always the concentration of SO_2 within category P_0 or Cl within category S_0 can be considered insignificant from the point of view of corrosive attack on carbon steel.

REFERENCES

[1] Leygraf, C; Graedel, TE. *Atmospheric Corrosion*, John Wiley and Sons: New York, NY, 2000, 9-11, 281-340.

[2] Guttman, H. *In Atmospheric Corrosion*, Ailor, W. H; Ed; John Wiley & Sons: New York, NY, 1982, 51-68.

[3] Mikhailovsky, YN. In *Atmospheric Corrosion*, WH. Ailor, Ed; John Wiley & Sons: New York, NY, 1982, 85-106.

[4] Johansson, LG; Vannerberg, NG. *Werkst. Korros*. 1981, 32, 265-268.

[5] Barton, K. *Protection against Atmospheric Corrosion*, John Wiley and Sons: London, 1976, 64-68.

[6] Vernon, WHJ. *Trans. Faraday Soc*., 1931, 27, 255-277.

[7] Tomashov, ND. *Theory of Corrosion and Protection of Metals*, McMillan, New York, NY, 1966, 454-481.

[8] Akimov, AG; Rozenfeld, IL. *Zashch. Metall*, 1978, 14, 445-449.

[9] Evans, UR; Taylor, CA. *British Corrosion J*, 1974, 9, 26.

[10] Walton, JR; Johnson, JB; Wood, GC. *British Corrosion J*, 1982, 17, 65-70.

[11] Ayllon, ES; Rosales, BM. In Anáis 14° *Seminário Nacional de Corrosão*, ABRACO: São Paulo, 1987, 26-34.

[12] Espada, L; Vasquez, ME; Sanchez, A. In Proceedings of the 13th *International Corrosion Congress*, ICC: Melbourne, 1996, Paper 010.

[13] Brown, PW; Masters, LW. *In Atmospheric Corrosion*, Ailor, W. H; Ed; John Wiley & Sons: New York, NY, 1982, 31-50.

[14] Tidblad, J; Mikhailov, AA; Kucera, V. Acid Deposition Effects on Materials in Subtropical and Tropical Climates. Data Compilation and Temperate Climate Comparison, KI Report 2000: 8E; *Swedish Corrosion Institute: Stockholm*, 2000, 33.

[15] Schweitzer, PA. *Atmospheric Degradation and Corrosion Control*, Marcel Dekker: New York, NY, 1999, 5-13.

[16] Kucera, V; Mattsson, E. *In Corrosion Mechanisms*, F. Mansfeld, Ed; Marcel Dekker: New York, NY, 1987, 211-284.

[17] ISO 9223. *Corrosion of Metals and Alloys -Corrosivity of atmospheres- Classification*, ISO: Geneva, 1992.

[18] Mikhailov, AA; Tidblad, J; Kucera, V. *Zashch. Met*, 2004, 40, 541-550.

[19] Corvo, F; Perez, T; Dzib, LR; Martin, Y; Castañeda, A; Gonzalez, E; Perez, J. *Corr. Sci*., 2008, 50, 220-230.

[20] Lan, TTN; Thoa, NTP; Nishimura, R; Tsujino, Y; Yokoi, M; Maeda, Y. *Corr. Sci*., 2006, 48, 179-192.

[21] Corvo, F; Perez, T; Martin, Y; Reyes, J; Dzib, LR; Gonzalez-Sanchez, J; Castaneda, A. *Corr. Sci*., 2008, 50, 206-219.

[22] Corvo, F; Haces, C; Betancourt, N; Maldonado, L; Veleva, L; Echeverría, M.; de Rincón, OT.; Rincón, A. *Corr. Sci*., 1997, 39, 823-833.

[23] Sica, YC; Kenny, ED; Portella, KF; Campos Filho, D. *J. Braz. Chem. Soc*., 2007, 18, 153-166.

[24] Tidblad, J; Mikhailov, AA; Kucera, V. *In Proceedings of the 14th International Corrosion Congress*, ICC: Cape Town, South Africa, 1999, Paper 337.2, 1-10.

[25] Veleva, L; Alpuche-Aviles, A. *In Outdoor Atmospheric Corrosion*, STP 1421; Towsend, E. T; Ed; ASTM: West Conshohocken, PA, 2002; 48-58.
[26] Oesch, S. *Corr. Sci*., 1996, 38, 1357-1368.
[27] Morcillo, M; Almeida, E; Rosales, B; Urruchurtu, J; Marrocos, M. Corrosión y Protección de Metales en las Atmósferas de Iberoamérica, Parte I: *Mapas de Iberoamérica de Corrosividad Atmosférica*, CYTED: Madrid, 1998, 473-685.
[28] Lien, LTH; San, PT. *In Outdoor Atmospheric Corrosion*, STP 1421; Towsend, E. T; Ed; ASTM: West Conshohocken, PA, 2002, 103-108.
[29] Misawa, T; Hashimoto, K; Shimodaira, S. *Corr. Sci*., 1974, 14, 131-149.
[30] Stratmann, M; Bohnenkamp, K; Engell, WJ. *Corr. Sci*., 1983, 23, 969-985.
[31] Kamimura, T; Hara, S; Miyuki, H; Yamashita, M; Uchida, H. *Corr. Sci*., 2006, 48, 2799-2812.
[32] Dillmann, P; Mazaudier, F; Hoerl, S. *Corr. Sci*., 2004, 46, 1401-1429.
[33] Castaño, JG; Botero, CA; Restrepo, A; Agudelo, EA; Correa, E; Echeverría, F. *Corr. Sci*., 2010, 52, 216-223.
[34] Ma, Y; Li, Y; Wang, F. *Mater. Chem. Phys*., 2008, 112, 844-852.
[35] Asami, K; Kikuchi, M. *Corr. Sci*., 2003, 45, 2671-2688.
[36] Slunder, CJ; Boyd, WK. Zinc: *Its Corrosion Resistance*, ILZRO: New Cork, NY, 1983, 1-112.
[37] Porter, F. *Corrosion Resistance of Zinc and Zinc Alloys*, Marcel Dekker: New York, NY, 1994, 1-262.
[38] Graedel, TE. *J. Electrochem. Soc*., 1989, 136, 193C-203C.
[39] Zhang, XG. *Corrosion and Electrochemistry of Zinc*, Plenum Press: New York, NY, 1996, 157-278.
[40] Castaño, JG; Arroyave, C; Morcillo, M. *J. Mater. Sci*., 2007, 42, 9654-9662.
[41] Odnevall, I; Leygraf, C. *In Atmospheric Corrosion*, WW; Kirk, H. Lawson, Eds; ASTM STP 1239; American Society for Testing and Materials: Philadelphia, PA, 1995, 215-229.
[42] Odnevall, I; Leygraf, C. *Corros. Sci*., 1994, 36, 1077-1087.
[43] de la Fuente, D; Castaño, JG; Morcillo, M. *Corros. Sci*., 2007, 49, 1420-1436.
[44] Morales, J; Martin-Krijer, S; Díaz, F; Hernández-Borges, J; Gonzalez, S. *Corr. Sci*., 2005, 47, 2005-2019.
[45] Morcillo, M; Almeida, E; Chico, B; de la Fuente, D. *In Outdoor Atmospheric Corrosion*, STP 1421, ET. Towsend, Ed; ASTM: West Conshohocken, PA, 2002, 59-72.
[46] Tidblad, J; Kucera, V; Mikhailov, AA; Knotkova, D. *In Outdoor Atmospheric Corrosion*, STP 1421, ET. Towsend, Ed; ASTM: West Conshohocken, PA, 2002, 73-87.
[47] Knotkova, D; Kucera, V; Dean, SW; Boschek, P. *In Outdoor Atmospheric Corrosion*, STP 1421, ET. Towsend, Ed; ASTM: West Conshohocken, PA, 2002, 109-124.
[48] ASTM G-50. *Standard Practice for Conducting Atmospheric Corrosion Tests on Metals*, ASTM: West Conshohocken, PA, 2003.
[49] ASTM D-2010. *Standard Test Methods for Evaluation of Total Sulfation Activity in the Atmosphere by the Lead Dioxide Technique*, ASTM: West Conshohocken, PA, 2004.
[50] ASTM G-140. *Standard Test Method for Determining Atmospheric Chloride Deposition Rate by Wet Candle Method*, ASTM: West Conshohocken, PA, 2002.

[51] ASTM G-1. *Standard Practice for Preparing, Cleaning, and Evaluating Corrosion Test Specimens*, ASTM: West Conshohocken, PA, 1999.
[52] Delgado, J; Castaño, JG; Correa, E; Restrepo, A; Echeverria, F. Rev. Fac. Ing. Univ. *Antioquia*, 2009, 50, 41-50.
[53] Gujarati, DN. *Basic Econometrics*, McGraw-Hill, New York, NY, 2003, 433-466.

In: Relative Humidity: Sensors, Management...
Editor: John M. Castillo
ISBN: 978-1-61761-734-8

Chapter 8

REVIEW OF RELATIVE HUMIDITY MANAGEMENT AND ITS ENVIRONMENTAL EFFECTS

José A. Orosa
University of A Coruña, Department of Energy
and Marine Propulsion, A Coruña, Spain

ABSTRACT

It is understood that indoor air must be classified and analysed based on different criteria like work risk prevention, energy saving, thermal comfort, biological sources, materials conservation and therapeutic effect. This study, in accordance with this classification, shows the indoor conditions that exist in different indoor ambiences, such as industrial environments, offices and school buildings, flats, libraries and a spa as well, using dataloggers that sample indoor conditions for long periods of time. Results revealed different limits of temperature and relative humidity, in concurrence with these criteria, and consequently, different procedures to control indoor ambiences. It is proposed, in particular, to define good behaviour habits in flats and libraries, like open windows for some time spans, to prevent the generation of excess humidity and improve the effect of passive methods like thermal inertia and permeable internal coverings in offices and schools. Hot ambiences especially, reveal two different control processes. On the one hand, in a spa, a low level of air change is proposed for energy saving, and low air velocity to prevent local thermal discomfort over wet skin surfaces. On the other hand, in industrial ambiences, it is proposed to increase the degree of air changes and enforce resting periods in a controlled ambience, to prevent work risk.

1. INTRODUCTION TO TEMPERATURE AND RELATIVE HUMIDITY CONTROL METHODS

Temperature and relative humidity are the chief parameters that define thermal comfort and influence energy saving in buildings. From this perspective, humidity problems can be identified in many applications, including office buildings, supermarkets, art galleries, museums, libraries, electronic manufacturing facilities, pharmaceutical clean rooms, indoor swimming pools and other commercial facilities.

Most of these problems surface, because in many HVAC applications, the cooling and dehumidification coil is unable to sufficiently meet the dehumidification requirements of the building when the latent load is high, either due to large internal moisture generation or through communication with the humid conditions outdoors.

To circumvent this problem, recent research papers [1] focussed on the optimal operation of building heating, ventilation, and air-conditioning (HVAC) systems, focusing on both temperature and humidity control. These new control systems attempt to minimize system power consumption while meeting building loads and maintaining comfort. Further, in these research works temperature and relative humidity parameters are also optimized within standard comfort constraints for materials conservation and health effects. The conclusion ultimately arrived at was that minimum energy use typically occurs at low airflow rates, with indoor humidity levels below the upper comfort limit. With this concept in view, variable-air-volume (VAV) systems were analysed and compared with constant-air-volume (CAV) systems and fan-coil systems [2]. Results showed that these VAV systems reached a minimum energy saving of 11.5% despite the variations in different building characteristics, climatic conditions and multiple zoning of each building.

Simultaneously, when air change is analysed to control the temperature, the relative humidity and dehumidification process are also being analysed. Researches like [3] revealed that employing hybrid systems with desiccant wheel results in remarkable savings in operating and higher plant costs with a notable reduction in the electric power demand and a better control of the ambient humidity.

This study shows the effect of temperature and relative humidity on the different types of indoor ambiences of ships, office buildings and flats, libraries and in a spa as well. In agreement with these criteria, different limits of temperature and relative humidity are seen, and consequently, different procedures to control indoor ambiences.

2. Materials and Methods to Study the Relative Humidity and Its Effects

As discussed earlier, different aspects of moist air need to be considered to predict its effect on energy consumption, comfort and materials conservation in buildings. In all the research works of this paper, temperature and relative humidity dataloggers were employed. These dataloggers [4, 5 and 6] permitted the researcher to sample indoor ambience variables of temperature and relative humidity, without interfering with indoor conditions for time spans as long as months or even for years.

Tinytag dataloggers

(Courtesy TinyTag)

PCE Datalogger

(Courtesy of PCE Ibérica)

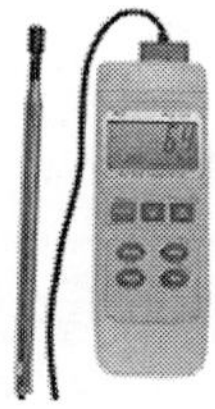

Portable Air velocity transducer PCE-009

(Courtesy of PCE Ibérica)

Figure 1. Portable temperature, relative humidity and air velocity dataloggers

Portable CO_2 IAQ910 transducer

(Courtesy of PCE Ibérica)

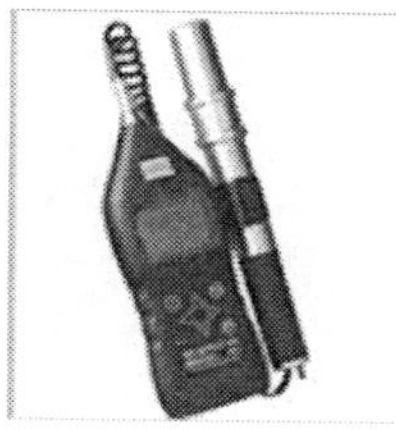

Casella Microdust

(Courtesy of Casella Ltd)

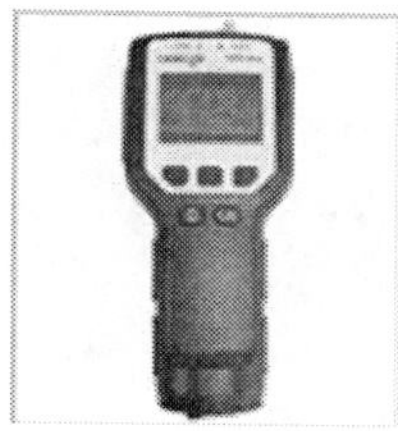

Casella VOC Pro Real

(Courtesy of Casella Ltd)

Figure 2. Portable IAQ dataloggers

As the total memory, namely the number of recorded samples of these dataloggers is constant, indicating that the longer the sampling time the lower the sampling frequency. Consequently, the time frequency is to be selected to suit the particular characteristics of each effect analysed. For example, moisture sources could be sampled with a time frequency of seconds or minutes, while thermal and hygroscopic inertia must be sampled with time frequencies of minutes or hours.

Finally, these dataloggers need less space than the expensive thermal comfort modules, and therefore, could be located near the worker for long time periods, as seen in Figure 1.

For both temperature and relative humidity, in some situations, a portable sample is needed to sample the Indoor Air Quality. Specifically, CO_2 and dust dataloggers allow us to define the air change per hour (ach) and the air quality, as shown in Figure 2.

3. Results and Discussion

When the materials typically employed during moist air research were identified, different indoor ambiences could be analysed with respect to air temperature and relative humidity conditions to define better indoor air control methods.

3.1. Indoor Air and Health Effects: A Practical Case Study of Flats

To study the real relationship between indoor conditions of flats and health problems, 25 Spanish flats, where at least one of the occupants have suffered from or were suffering with a respiratory ailment at the time of the study, were selected. The aim was to relate the temperature and relative humidity that existed in the indoor ambience over a 24-hour period of household life, to the air renovation and microbiological load. To achieve this, the average humidity ratio relating to the whole set of flats was determined for the kitchen, living room, bedroom and the children's bedroom for the following time periods: breakfast, from 7:00 to 10:00, lunch, from 12:00 to 15:00, dinner, from 20:00 to 21:00 and sleeping time 12:00 to 7:00.

Results showed that indoor air temperature was above 20°C, for an outdoor average value of 12°C. The outdoor humidity was generally lower than that prevailing indoors, with an average humidity between 5.74 g/kg and 9.5 g/kg over the whole set of rooms and homesteads. The main result revealed that, with the exception of the kitchen at lunchtime, the humidity in the bedroom was the highest, than at the other periods of time.

Comfort conditions showed that the indoor ambiences in the bedroom during the night, and in the kitchen during lunchtime were within 15% of the percentage of local dissatisfied of local thermal comfort; therefore, they could be considered comfortable ambiences. Despite this, the ventilation rate, expressed as a function of the level of CO_2 concentration in the room, was abnormally elevated. Values higher than 3000 ppm have been obtained, particularly in the bedroom with the door closed, resulting in a lower ventilation rate than the accepted minimum. Further, an analysis was made of the influence between rooms, and the high influence between the kitchen and the living room during dinnertime.

Once the humidity level and its influence on the nearer rooms were analysed, fungal growth was researched. This study showed that the fungi are present at much lower concentrations in flats with lower air relative humidity. Further, no clear relationship was found between fungal levels and the number of occupants or pets. However, a clear relationship existed between the fungi and the damp on the walls.

The general conclusion drawn from this work revealed the need to foster behavioural habits of opening the outer doors and windows during periods of maximum humidity generation, particularly during the lunch and sleep hours.

3.2. Indoor Air and Energy Saving: A Practical Case Study of Offices and Schools Buildings

In the north west of Spain and Portugal, a region of humidity exists due to the humid winds blowing in from the sea. Consequently, the public buildings reveal the highest energy consumption by way of air conditioning during the winter. During spring however, the heating system is employed only if the indoor conditions reach certain temperatures and relative humidity values. To improve these ambiences and reduce the energy consumption, passive methods like permeable coverings and thermal inertia can be utilized. These are the new theories being recently analysed in depth. These permeable coverings minimise the relative humidity peaks, particularly during the first hours of occupation [7, 8]. Thermal inertia in the walls helps to store energy in times of abundance and release it when the indoor temperature drops.

To define this effect in real buildings, indoor conditions in two school buildings simulated with high and low constructional thermal inertia, respectively, were sampled. The old school buildings revealed high wall thermal inertia due to being constructed with stone and concrete on the inner wall surfaces. The new buildings however, revealed low constructional thermal inertia, where the walls show layers of insulation, brick, concrete and plaster arranged symmetrically, with respect to the middle of the wall.

As the objective of this study was to define the effect of thermal inertia on the indoor and outdoor conditions, the variables of temperature and relative humidity were sampled for a whole year. Next, based on this real sample data, the HAM-tools software building model [9-14], was modified. After good adjustment between the real data and simulated results, the effects of ventilation rate, internal coverings and orientation over global thermal inertia, and thus over the energy consumption could be predicted. Simulations of this research work showed that old buildings show lower thermal inertia than the new ones because of solar heat gain. Therefore, a building with a lower time constant reacts faster to weather changes and variations in internal heat gains than the more heavy buildings. Parameters like air changes and permeable coverings interact similarly, over a time constant. If the air changes are reduced, an old building will experience a slow increase of the time constant, while simultaneously, a new building will experience a clear rise in internal temperature due to heat gain. Finally, the effect of permeable coverings on indoor ambiences is reduced if very little permeable surface coverings are present.

The general conclusion drawn from this work is that simulation software resources facilitate predictions during the process of designing buildings and enable corrections of real buildings to be easily proposed. It is thus very interesting to validate this resource with real sampled data.

3.3. Indoor Air and Materials Conservation: A Practical Case Study of Libraries

Another research work was developed to relate indoor temperature and relative humidity with materials conservation. A practical case study on indoor conditions in a Spanish naturally ventilated library was conducted to prove this theory. As earlier mentioned, the northwest of Spain enjoys high relative humidity throughout the year. This outdoor relative humidity reaches mean values of 88% at temperatures of 17°C during summer months. The months of June and July were specifically chosen as higher levels of occupation are noted then, and consequently, indoor conditions under natural ventilation reach their maximum limit values.

Environmental limit levels are based on ASHRAE indications [15] and Environmental Specifications for the Storage of Library & Archival Materials [16] whose recommended environmental ranges are listed for books, paper, and people. These specifications showed that for general stacks and branch libraries, temperatures of 20-22°C with relative humidity between 40-55% are recommended. As rapid fluctuations damage collections the most, and the aim of using this ambience is to maintain constancy within the specified ranges for temperature and humidity, while minimizing daily and weekly variations.

When indoor comfort conditions with respect to sampled data were compared, the indoor temperature in the library did not show a daily swing reaching a mean value of 22°C, during this two-month period. In the archives, the indoor temperature showed a higher mean value of 25 °C, during the months of June and July, and clear daily temperature changes varying with the library occupation. This indicates that the temperature exceeds comfort limits and materials conservation, resulting in paper fragility, particularly with humidity change. The reason for the differences between these two ambiences was related to the fact that the library had a porch that reduced the radiative heat gain. This, coupled with low air renovation, is the cause for this temperature difference.

Results showed that in the library, despite natural ventilation reducing the temperature peaks, the heat and moisture gains resulting from the occupants reach the archive downstairs, and consequently affect book preservation.

In the archive, an approach to materials preservation was done. In this sense, indoor air relative humidity revealed a mean value of 58%, at a mean temperature of 25°C. Although this relative humidity was not as high as that of the library, the high temperature led to mould growth that was detected on the wall constructions in book damage. It was related to the high relative humidity over a long period of time, a lower air speed and the clear presence of dust.

After and structural and thermodynamic analyses of the building, the conclusion drawn was that this archive revealed a higher temperature mean value due to the low external air renovation and the existent solar radiant gain that affected some of its walls. However, as earlier explained, heat and moisture sources alone are not responsible.

Therefore, as a possible solution for materials conservation, adequate communication between zones by providing doors and employing other passive methods to reduce archive temperature, such as wall thermal inertia, is done underground.

3.4. Indoor Air and Work Risk: A Practical Case Study of Industrial Environment

The thermal parameters of indoor environments call for a suitable study; knowledge and processing in relation to industrial safety, because high or low air temperatures and uncontrolled heat sources may induce lower productivity rates, higher accident rates and greater health hazards. A ship is good example of a thermal indoor environment with extremely changing characteristics in short intervals of time. With this in view, sampling of air temperature and relative humidity was performed in a merchant ship.

In a ship, the engine room possesses the characteristic ambience that must be analysed in depth, especially because of the special characteristics related to work safety and heat sources. Temperature and relative humidity data from the engine room and other locations were analysed to obtain comfort indices using the new models applied to this environment.

Further, given the influence of outdoor conditions on indoor environment, outdoor data was collected during the monitoring months from December to February. The results showed the outdoor conditions of mean temperature 23°C, with a maximum value of 27°C, and an average relative humidity of 60%. The weather was predominantly sunny and not too cloudy.

Indoor environments were assessed based on the study of comfort indices; however, the variables involved in this definition must remain within the standard range provided. Thus, a statistical analysis of temperature and relative humidity data in each indoor zone was done. Results showed that the average temperature of the air both on the bridge and in the dining room remained at about 22°C. In the engine room, however, the average was 32.5°C, with peaks of 38.5°C. These results are outside the values permitted for hot environments and can precipitate various health disorders.

From the general conclusion drawn from the analysis of these real conditions, some indications were proposed. The first was an increase in air changes with outdoor air to control the temperature and relative humidity of these ambiences. The second was to define the limit time that a person could work without heat stress, concurring with Spanish standards NTP, as well as to consider the following work risk-prevention measures:

- Drinking water. Sources of drinking water must be available close to work locations, and workers need to be educated on the necessity of drinking frequently.
- Acclimatization. Workers starting new jobs or returning to work require an exposure time to achieve acclimatization.
- Metabolic heat. Adjusting the length and frequency of breaks and work periods and work rates may reduce the metabolic heat release. As far as possible, work must be scheduled during times of less heat. Work periods in the engine room must not exceed 27minutes, after which the worker must remain for about 10 minutes in the control room.
- Workers must be kept under constant surveillance by a trained colleague to detecting any symptom of heat strain.

3.5. Indoor Air and Its Therapeutic Effect: A Practical Case Study of a Spa

A spa is an interesting example of new indoor ambience, and therefore, an ambience that currently presents new problems of temperature and relative humidity control. These ambiences reveal pools of varying water temperatures, aiming to relax the body musculature and thereby reduce health problems. This has certain effects on energy consumption and materials degradation.

Water temperatures typically hover around 25°C, while air temperatures of 27°C are commonly obtained. According to the Spanish national standards, the surrounding air must range between 2 and 3°C to reduce the water evaporation and thus, the energy consumption. When analysing the water temperature in a pool, it must be remembered that all components of seawater can be absorbed through the skin in 20 minutes when the water bath is at a temperature of 37°C. Consequently, private pools should be maintained at a temperature of 26-27°C although, most often, they are requested to maintain a temperature of 30°C.

The reason that these temperatures are proposed is based on the thermal comfort of the occupants, to prevent the wet skin from emanating a higher heat release from the client's body. Another reason is that a distinct relationship exists between the evaporated water and the temperature and the relative humidity of the air. By the same standard, an indoor air relative humidity below 60 to 65% is proposed to prevent condensation on the cold surfaces in contact with the outdoors, like walls or windows.

Hourly air changes in the different zones of the spa must be studied. Particularly for pools, the Spanish national standard has proposed a minimum ventilation rate of 12.5 l/s person, 1.2 decipols and a maximum CO_2 concentration of 500 ppm. Another indication is the minimum air flow of 2.5 l/s m^2 of pool surface that adds an air flow to control the relative humidity. Finally, pools must maintain 20 to 40 Pa under pressure with respect to the closer rooms.

This research work aims to analyse the relationship between indoor air parameters of temperature and relative humidity and health effects, thermal comfort, energy saving and materials conservation while preventing condensations. Therefore, a Spanish spa that employs sea water as pool fluid was analysed. This particular ambience was selected as this fluid causes specific corrosive problems anywhere. The ISO 7730 [17] and the NTP 689 [18] public swimming pool (I) were used as reference. Risks and prevention and ASHRAE indications were noted. These standards indicated that the relative humidity must be maintained between 50% and 70%, preferably at 60%, to reduce the water lost by evaporation and energy expenditure for treatment of outdoor air. However, higher values could precipitate condensation problems, due to humidity, on the walls and ceilings.

When these indications were compared with the sampled data, the heating and air humidity in general and in Thalasso pools was observed to have values reaching upto 74.9%.The pool area, however, showed very high temperature values (31.5°C) in any of the measurements. Therefore, as the first control method for this high relative humidity, the temperature is recommended to be reduced. More typically, considerations for control methods of these special ambiences are:

1. To prevent energy loss it is usually proposed to increase the relative humidity of the enclosures, thus reducing evaporation and to avoid condensation on the walls.
2. Air velocity within the enclosure needs to be controlled, as it is another parameter related to evaporation, and therefore, with energy loss in the installation.
3. Air changes have proved to be a critical parameter in energy consumption throughout the system. If the air change in the air is doubled, the power consumption is doubled.
4. Finally, thermal inertia exerts a minimal influence on the overall consumption of the building.

On analysing these general indications, other specific indications to control indoor temperatures and relative humidity in a spa must also be considered. For example, the new Spanish standard RITE [19] prohibits the use of non-renewable energy for air conditioning of both indoor and outdoor swimming pools, solar energy being the mandatory source to be used, for 30 to 70% of the annual heat requirements for heating the pool. However, water heating for outdoor swimming pools may be done using renewable energy sources (solar energy) and waste (biomass) only, although heat pump equipment is permitted. The use of conventional energy is strictly forbidden. Finally, when the pool is not in use, the water layer must be protected with a thermal barrier.

The last parameter to be analysed in such types of indoor ambiences is the local thermal comfort. In this case, 30% dissatisfaction in the lower floor and 80% in pools were obtained. This was due to the high temperature and relative humidity in the pool. Thus, a person can only stay in this kind of ambience during limited periods of time, to reduce mucous drying risk. Finally, sudden temperature changes from the zones of high temperature to the halls must be considered. A rapid change of temperature was observed, particularly in the corridor areas.

CONCLUSIONS

General conclusions can be drawn from this research work. Temperature and relative humidity together with air changes, clearly influence comfort conditions, energy saving and materials conservation.

Methods of temperature and relative humidity control depend on the particular effect that one desires to control. For example, if energy consumption in office buildings is to be controlled, the passive methods of controlling thermal and hygroscopic inertia must be improved. Despite this, parameters like building orientation and occupation habits could alter these conditions, as HAM-tools reveal. However, when moisture sources are to be controlled their origins need to be considered. For example, if the origin is a punctual generation during cooking time, during the night or from a shower, the more economical control method would be by natural or forced ventilation that permits the reduction of excess humidity ratio in the building. This excessive indoor humidity level resulted in fungal growth, and consequently, with health problems.

If the origin of the moisture source is continuous and released to the air in high amounts, new methods apart from air temperature, relative humidity and air changes must be considered, to attain improved ambience. In this last instance, a spa represents a special case that must be seriously studied, as the high relative humidity and temperature generated can culminate in moisture condensation on the surfaces in contact with the outdoors. Further, their special ambience requires a deeper consideration of the thermal ambiences of each indoor environment, to protect the clients from thermal shock, and control the draught over across the wet surfaces.

Finally, extreme industrial ambiences like the engine room of a merchant ship reveals high temperature and relative humidity that call for work risk prevention measures, particularly with respect to dehydration. These kinds of ambiences can only be controlled with adequate air changes, and their occupants can only release the excess heat in control zones like the control engine room.

As a general conclusion, temperature and relative humidity are confirmed as the main parameters to be considered when controlling indoor ambience. New control techniques for heating ventilation and air conditioning systems (HVAC) that accommodate all these different aspects, must therefore be developed [20].

ACKNOWLEDGMENTS

I would like to thank the University of A Coruña for sponsoring project 5230252906.541A.64902.

References

[1] Huh, Jung-Ho; Brandemuehl, MJ. Optimization of air –conditioning system operating strategies for hot and humid climates. *Energy and Buildings*., 2008, 40, 1202-1213.

[2] Yao, Y; Lian, Z; Liu, W; Hou, Z; Wu, M. Evaluation program for the energy-saving of variable-air-volume systems. *Energy and Buildings*, 2007, 39, 558-568.

[3] Mazzei, P; Minichiello, F; Palma, D. HVAC dehumidification system for thermal comfort: a critical review. *Applied Thermal Engineering*, 2005, 25, 677-707.

[4] PCEibérica http://www.pce-iberica.es/ (Accessed February 2010)

[5] Tinytag http://www.geminidataloggers.com/ (Accessed February 2010)

[6] Casella http://casella-es.com/ (Accessed February 2010).

[7] Orosa, JA; Baaliña, A. Passive climate control in Spanish office buildings for long periods of time, *Building and Environment*, 2007. doi:10.1016/j.buildenv.2007.12.001

[8] Orosa, JA; Baaliña, A. Improving PAQ and comfort conditions in Spanish office buildings with passive climate control, *Building and Environment*, 2008, doi:10.1016/j.buildenv.2008.04.013.

[9] Weitzmann, P; Sasic Kalagasidis, A; Rammer Nielsen, T; Peuhkuri, R; Hagentoft, C. *Presentation of the international building physics toolbox for Simulink*. 2009.

[10] Kalagasidis, AS. BFTools H Building physiscs toolbox block documentation. Department of Building Physics. *Chalmer Institute of Technology*, Sweeden. 2002.

[11] Kalagasidis, AS. HAM-Tools. *International Building Physics Toolbox*. Block documentation.

[12] Weitzmann, P; Kalagasidis, AS; Nielsen, TR; Peuhkuri, R; Hagentoft, C. *Presentation of the international building physics toolbox for simulink*.

[13] Nielsen, TR; Peuhkuri, R; Weitzmann, P; Gudum, C. Modelling Building Physics in Simulink. BYG DTU Sr-02-03. ISSN 1601-8605. 2002.

[14] Rode, C; Gudum, Weitzmann, P; Peuhkuri, R; Nielsen, TR; Sasic Kalagasidis, A; Hagentoft, CE. International Building Physics Toolbox-General Report. Department of Building Physics. *Chalmer Institute of Technology*. Sweden. Report R-02: 2002. 4.

[15] ASHRAE. 2009. Fundamentals. Atlanta. American Society of Heating, *Refrigeration and Air Conditioning Engineers*, Inc.

[16] Environmental Specifications for the Storage of Library & Archival Materials. http://www.solinet.net/emplibfile.(Accessed February 2010)

[17] ISO 7730:2005. *Ergonomics of the thermal environment- Analytical determination and interpretation of thermal comfort using calculation of the PMV and PPD indices and local thermal comfort criteria*.

[18] NTP, 689 public swimming pool (I). *Risks and prevention*.

[19] RITE, 2007. Spanish Standard for Thermal Installations in Buildings.

[20] Orosa, JA. University of A Coruña. *Procedimiento de obtención de las condiciones de temperatura y humedad relativa de ambientes interiores para la optimización del comfort térmico y el ahorro energético en la climatización*. Patent: P200801036.

In: Relative Humidity: Sensors, Management...
Editor: John M. Castillo

ISBN: 978-1-61761-734-8

Chapter 9

Relative Humidity: Sensors, Management and Environmental Effects in a Spa

José A. Orosa*

University of A Coruña, Department of Energy and Marine Propulsion, A Coruña, Spain

Abstract

This chapter covers the indoor conditions of temperature and relative humidity in the special indoor ambience of a spa. The effect of these last two parameters, in particular, thermal comfort and energy saving will be analysed, revealing the main tendencies in accordance with the prescribed standards. Results showed that indoor air temperature depends on the pool water temperature, and that this in turn depends of the kind of health effect that we are looking at. Consequently, energy optimization is limited to spa application. However, the percentage of dissatisfied clients and indoor air acceptability showed a certain local discomfort in this ambience; therefore, a low residence time was required. Finally, some energy saving procedures were proposed to control the relative humidity of the ambience in the pool, like utilizing a protective covering over the pool, when not in use. Research works in future were proposed to control all these parameters based on the new systems to control temperature and relative humidity as well as new software resources.

Keywords: Spa, Thalasso, ISO 7730, RITE, NTP 689.

1. Introduction

According to the latest research results [1], the combination of spa treatment and physical exercise decreases the symptoms of pain and blood pressure. Besides, these studies have indicated that the combined program decreases the intensity of pain in arthritic patients and

* Corresponding author: E-mail: jaorosa@udc.es

does not cause any complications in the elderly or in neurology patients. Another interesting result from this study was that no clear differences exist between men and women on the effects of the spa.

Once the beneficial effects of spa treatment on health are recognized, analysing the origin of this analgesic effect gains interest. It is particularly interesting to relate the resultant health effects to the characteristics of water as in the research work conducted by Ledo [2]. He dealt with different Spanish spas and their effects resulting from the water used in each.

From among the different types of water employed in each spa, Thalasso therapy emerges of special interest as this treatment uses sea water [3]. According to Ledo [2] this type of therapy combines the effects of exposure to seawater ambience, i.e. the mineral water content and sunbathing. Thalasso therapy aims at absorbing all the elements found in seawater in high concentration, particularly iodine, calcium and magnesium. To achieve this, all these components can be absorbed in a 20-minute bath, at a water temperature of 37ºC. In this context, the water temperature ranges, interestingly, supported by men, are shown in Table 1.

Table 1. Water temperature bearable by humans

Temperature	Effect
Very cold (0-12ºC)	Pain Sensation
Cold (12-18ºC)	Sensation of cold; Poor tolerance
Slightly chilled (18-27°C)	Feels nice, short exposures.
Neutral (27-32ºC)	Feels nice
Temperate (32-36.5°C)	Feels comfortable
Heat (37-40ºC)	Heat sensation
Very warm (40-43°C)	Tolerable in short exposures

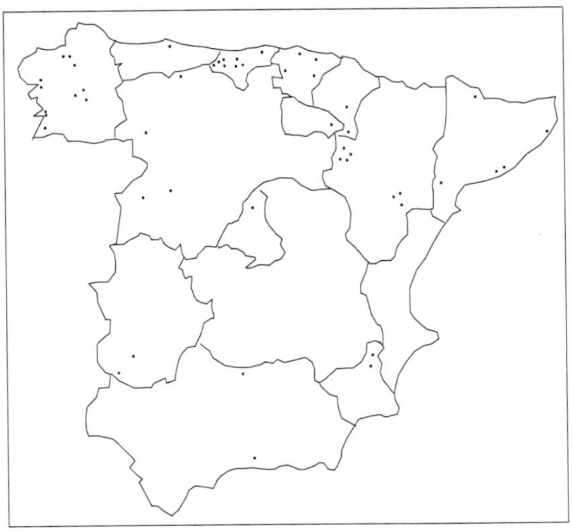

Figure 1. Location of Spanish spa

Another reason for appreciating seawater as a spa fluid is related to its low susceptibility to radioactive contamination [4]. This is an interesting concept as certain Spanish regions reveal spa water with Radon measurements of 824 Bq/l. A high percentage of spas revealed specific levels of Radon contamination in north-western Spain, as seen in Figure 1. This concentration varies with each spa centre in the range from <4 Bq/l to 840 Bq/l, while in the air it reaches to over 5000 Bq/m^3. Further, it poses risks not only to swimmers but to spa workers as well, as they too are exposed to the same ambience, which is recognized as an occupational hazard.

Once it became evident that sea water as a bathing fluid in a spa was gaining popularity, more studies on energy consumption, materials conservation and comfort conditions [5-9] were recently done. Recent studies particularly on the energy consumption in a spa showed [6] that about 75% of the total heat load is invested in the processes of evaporation and ventilation that occurs in the pool area.

A typical solution for this evaporation problem is to increase the relative humidity of pool ambience, and thus reduce the evaporation. However, this solution is not always feasible and reveals limitations, given the possibility of moist air condensing on the walls of the premises, if the relative humidity exceeded 50%. Values of air temperature and renewal that are of special interest in these applications are listed in Tables 2 and 3.

Once energy saving regarding pool water evaporation control was proposed, some research works on modelling of moisture loss in pools were done [10, 11]. Such studies proposed equations that indicate the percentage of water loss (g/h) in pools by evaporation, according to equation 1.

Table 2. Indoor conditions of a spa during the winter

Winter indoor conditions	
Maximum project temperature	24°C
Minimum project temperature	18°C
Mean project temperature	22°C
Relative humidity	<70%
Temperature according to type premises (winter)	
Dressing rooms	20°C
Toilets	20°C
Services area	20°C
Reception and offices	18°C
Medical offices	20°C
Hall	20°C
Treatment room	24°C
Solarium and pools area	24°C
Lobbies and corridors	18°C
First-aid room	20°C
Gyms	18°C

Table 3. Indoor conditions of a spa during the summer

Temperature according to premises type in summer		Air ventilation requirements (dm^3/s)			
		Per person		Per m^2 surface	
		Minimum	Maximum	Minimum	Maximum
Dressing rooms	24°C	15	22	2.5	5
Toilets	24°C	-	-	2	3.5
Services area	24°C	7	12	-	-
Reception and offices	23°C	7	10	0.5	-
Medical offices	24°C	7	12	-	-
Waiting room	24°C	5	8	-	-
Treatment room	27°C	7	12	-	-
Solarium and pools area	27C	10	14	-	-
First aid-room	24°C	7	12	-	-
Gyms	23°C	10	14	-	-
Summer indoor conditions					
Maximum project temperature	27°C				
Minimal project temperature	23°C				
Half project temperature	24.5°C				
Relative humidity	60%				

$$M = [816 + 133 \cdot n) \cdot (W - RH \cdot W)] \cdot S + 100 \cdot n \cdot S \quad (1)$$

Where
M is the maximum evaporation in (g/h).
S is the pool surface in (m^2).
n is the pool occupation.
W grams of water per kg dry air (g/kg).
RH is relative humidity (%).

In this equation, a value of 0.2 for occupation is experimentally proposed. In addition, a ratio of 100 g/h it indicates the amount of water vapour emitted per hour per person. Finally, in this example, the conditions in the spa are assumed to remain constant during the night. To validate this model it observed that, assuming an occupancy of 1 h/day and n= 0.2, a maximum evaporation of M=11.21 l/h will result with a pool water consumption of 39.22 l/day.

The results of these studies on pool water evaporation reveal curves where the evaporation per square meter and hour of swimming increases with temperature, relative humidity and surrounding air speed. Therefore, controlling the air velocity in this enclosure is an added incentive, as it is another parameter related to the energy lost in the installation.

Other studies discussed new energy saving methods in the spa, particularly the possibility of energy utilization of the heat of waste water for air conditioning applications [8, 9].

After analysing the latest research works regarding this type of ambience and identifying the advantages, disadvantages and options for air conditioning and energy use, an analysis was done of our particular spa. In the present chapter, a discussion was conducted aiming at exposing the energy saving methods employed in such facilities, in accordance with the RITE, ISO and NTP standards [12-17].

2. Materials and Methods

2.1. Building

In this chapter, the building under study is located in the northwest of Spain, which experiences high relative humidity over most of the year, due to the effect of the Atlantic Ocean winds, and a mild mean outdoor temperature, with an average maximum temperature of 25°C and a minimum of 3°C. In such climatic conditions, heating is used to control the indoor temperature during the cold season. In summer, temperatures soar, with natural ventilation being the most effective passive cooling strategy.

This building is a Spanish spa with a base area of 11,040 m^2 for all uses, for each of its three plants. It presents different zones and activities but principally, four pools with different sea water temperature and a gym are present.

2.2. Indoor Air Temperature and Relative Humidity

To analyse the effect of temperature and relative humidity in this special indoor ambience, dataloggers were used to take samples at different locations of the building. The indoor temperature and relative humidity particularly, in each pool, were related to water temperature and indoor relative humidity, concurring with national standards. Further, wall temperatures and air velocity were also considered. This sampling process, based on the ISO 7730 indications, was done at 5-minute intervals for each variable.

At the same time, the outdoor conditions were sampled to identify the effect of weather on indoor conditions. The outdoor conditions were sampled using weather stations located near the spa.

2.3. Local Thermal Comfort Models

To conduct an in-depth study of the indoor thermal comfort conditions in this ambience, the thermal comfort models developed by P.O. Fanger, in accordance with the ISO 7730 [12] and its equivalent ASHRAE [13] standard must be employed. As these standards are applied only to thermal conditions that, on average, affect the environment, they are termed general thermal comfort. To appreciate thermal comfort from a broader perspective, one needs to focus on the particular conditions involved or what is otherwise called local thermal comfort. Several empirical equations have been employed by some authors over the last few years to study indoor air quality. Specifically, local thermal comfort models must define the relationship between the air temperature and relative humidity conditions, and the perception of indoor air. Consequently, the indices of percentage of dissatisfied (PD) and Acceptability of cooling air (Acc) were employed. Therefore, the local function of equation 2, called Percent of Dissatisfied, obtained by Toftum *et al.,* [18, 19] was applied to estimate the effect of humidity on thermal comfort.

In this research, the response of 38 individuals, who were provided with clean air in a closed environment, was studied. The air temperature and relative humidity conditions ranged between from 20°C and 45% to 29°C and 70%, respectively. During the experiment, individuals assessed the ambient air with 3 or 4 puffs, and thus the equation for the percentage of local dissatisfaction was developed, as shown in equation 2. This equation is found to be applicable in the study of the spa conditions in line with these model conditions.

Before applying this index to analyse the indoor ambience, ASHRAE recommendations must be considered, keeping the percentage of local dissatisfaction below 15% and the percentage of general thermal comfort dissatisfaction below 10%. From this equation the PD index is observed to tend to decrease as the temperature decreases.

$$PD = \frac{100}{1+\exp(-3.58+0.18(30-t)+0.14(42.5-0.01P_v))} \tag{2}$$

Where t is temperature (°C) and P_v is the partial pressure, (Pa).

The perception of thermal comfort and local air quality are intimately interrelated and are strongly influenced by the effect of humidity. To estimate air acceptability, the Acceptability index function of equation 3 can be applied. This index is a function of the enthalpy, thus by increasing the enthalpy, Acceptability decreases.

$$Acc = aH + b \tag{3}$$

Where H is the enthalpy, kJ/kg, and a and b are constants.

3. Results and Discussion

Once indoor and outdoor conditions were sampled, some recommendations were proposed based on air temperature, the pool water temperature, and the soil temperature.

3.1. Pool Water and Air Temperature

As a clear relationship exists between water and air temperature in such indoor ambiences, the principal applicable standard recommendations will be discussed. In this instance, the prescribed Spanish standards indicate that the pool water temperature, except for therapeutic purposes, shall be maintained between 24 and 30°C. This temperature should vary with use, for example, with the metabolic activity of the bathers. For sports competition and training, water temperature should be maintained between 24 and 25°C, and one or two degrees higher for recreation, splash and teaching, as seen in Table 4. In private pools, the water temperature should be maintained at a temperature of 26-27°C, although they may often be requested to maintain a temperature of 30°C.

Despite water temperature being a fundamental parameter in such ambiences, the standards recommend that it must be homogeneously distributed throughout the pool, both horizontally and vertically, particularly with an error margin of ± 1.5°C. According to this norm, to avoid the unpleasant effect of temperature stratification, hot water should be injected into the pool from the bottom of the walls.

Spanish national standards recommend a mean temperature of 27°C, relative humidity of moist air of 65% and a water temperature of 24-32°C in pools, depending on the use of the pool. For example, as a general rule, air temperature is always advised to be maintained 2-3 degrees greater than that of the water and the relative humidity around 65%. There are two reasons for selecting the temperature and relative humidity limits. The first reason is to avoid the cold feeling of swimmers emerging wet at a low ambience temperature. Secondly, there is a direct relationship between evaporation from the pool water and the environmental temperature and humidity, with its corresponding energy loss. Further, the relative humidity must be kept between 60 to 65%, to avoid condensation on the walls in contact with the outside.

Before beginning an analysis of the indoor conditions of temperature and relative humidity, the outdoor conditions must be noted. These conditions reveal a mean temperature and relative humidity of 23.7°C and 58.6%, respectively, during the hours selected to sample indoor ambience.

If now we compare the real sampled data of Figures 2 and 3 with the standards mentioned earlier, it can be observed that the pool area shows high temperature, in any of the measurements. For example, the temperature and air humidity in Thalasso pools are above the standard limits, reaching a value of 31.5°C. Although the moist air temperature is fixed consistent with the water temperature which in turn varies with the specific therapy, the temperature of pool is recommended to be reduced.

The relative humidity value touches a maximum of 65% and 74.9% in the general pools and in Thalasso, respectively. These values exceed the standard indications, and consequently, may run the risk of condensation and thus, even higher energy consumption than the other pools. Despite this, the medical treatment of Thalasso therapy necessitates these conditions; therefore, only preventive methods to maintain this ambience can be proposed.

Table 4. Water temperatures S/RITE 10.2.1.2

Competition	24°C
Training	26°C
Education and Recreation	25°C
Physically handicapped	29°C
Children's pool	30°C
Children 3 to 6 years and seniors	32°C
Pregnant women	30-32°C

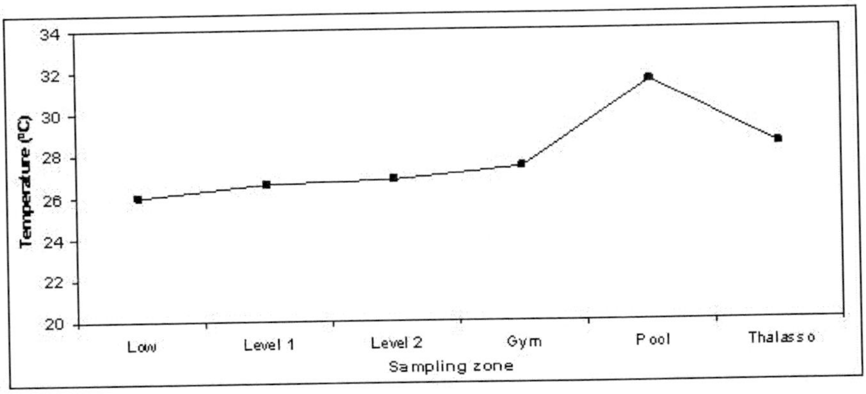

Figure 2. Temperature at the different zones of the spa

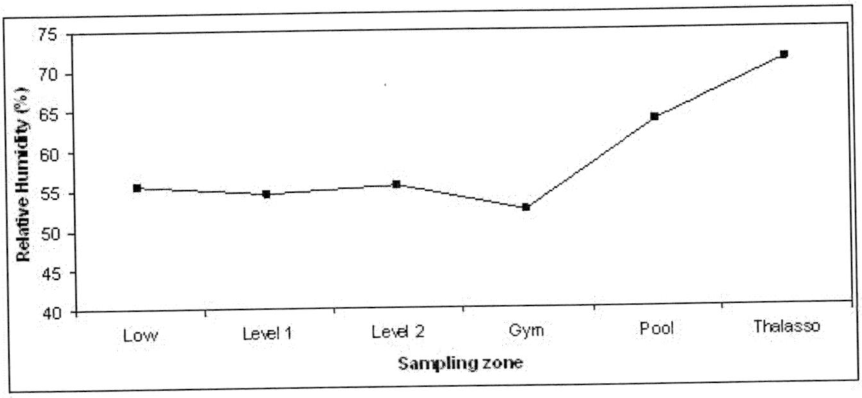

Figure 3. Relative Humidity at the different zones of the spa

Finally, as evident in Figure 2, the moist air relative humidity in other parts of the building reveals relative humidity similar to that of the outdoors, except in the gym. Here, low relative humidity is observed, as a result of specific treatment of its indoor air using individual Heating Ventilation and Air Conditioning systems. The reason the gym area is accorded special consideration is because of its indoor activity and therefore, the need for a high degree of air change of indoor air.

To control both indoor and outdoor swimming pool ambiences, the Spanish standard (RITE 2007) prohibits the use of non-renewable energy for water conditioning, solar energy being mandatory to provide between 30 and 70% of the annual heat requirements to warm the pool. A heat gain of about 10ºC above the ambient pool water temperature is expected. The water surface of both the outdoor and indoor heated pools must be covered [20].

In implementing this rule, the size of the solar collectors must be considered. The collectors used to normally weatherize a pool will depend of the m^3 of sea water utilised. This rule specifically indicates that the surface of the solar collector must be at least equal to 50% of the pool surface.

Table 5. Maximum floor temperature

Finishing	Temperature
Textiles	21°C
Wood	22-24 °C
Linoleum	24 °C
Concrete	26 °C

Another method to manage this special kind of indoor ambience is to control the energy exchange between the pool water and its surrounding moist air. For example, when the season starts, the pools take about three days to reach adequate working temperature. Throughout the season, both the water temperature and pool indoor air should be maintained constant, due to the large thermal inertia of the entire system. RITE insists that the water layer be protected with a thermal barrier when the pool is not in use. This obligation is mandatory in indoor pools and is highly recommended for outdoor pools, as renewable or waste energy alone is to be employed to heat this pool.

After analysing the required standard for water and moist air temperatures, the floor temperature in this particular ambience too must be considered. As earlier mentioned, occupants are in their bathing suits most of the time and therefore, the maximum floor temperature should be 28°C, while the minimum will vary, depending on the type of floor finish, as seen in Table 5. The surface temperature of the floor is to be as high as the heat conductivity of the floor finishing material used.

3.2. Perceived Air Quality and Expected Percentage of Dissatisfied

To study the air quality of these indoor ambiences in depth, the principal standard indications must be considered. RITE specifies a minimum ventilation air flow to be added, if necessary, to control the relative humidity. This standard shows a mean value for heated pools equal to 2.5 l/s per m^2 of water surface layer and ISO 13790:2004 proposes a minimum ventilation rate of 15 m^3/h person. Besides, this air must be supplied in a way that the pools kept in depression respect outdoors, between 20 and 40 Pa, with respect to the adjacent premises. Finally, the speed of air flow in the different sections of the Air Treatment Unit (ATU) is recommended to never exceed 2.5 to 2.6 m/s, in order to reduce internal pressure losses or exfiltrated drafts.

If these standards are applied to this case study, and if this building is assumed to have occupation peaks of 500 persons during the day, it can be assumed that a minimum air change of 7500 m^3/h will be needed to ensure1.2 decipoles and a maximum concentration of 500 ppm CO_2.

Once the indoor conditions were sampled and this ambience was proved to have an adequate air change, the perception of indoor air was analysed. To achieve this, the Percentage of Dissatisfied index (PD), was utilized as earlier mentioned. Percentage of Dissatisfied Index has proved to be acceptable in the low and in Level 1, while in the gym and, in a much more intense way in the pools, this value is observed to increase. Thus, we move from a 30% dissatisfied in the low to 80% in the pools.

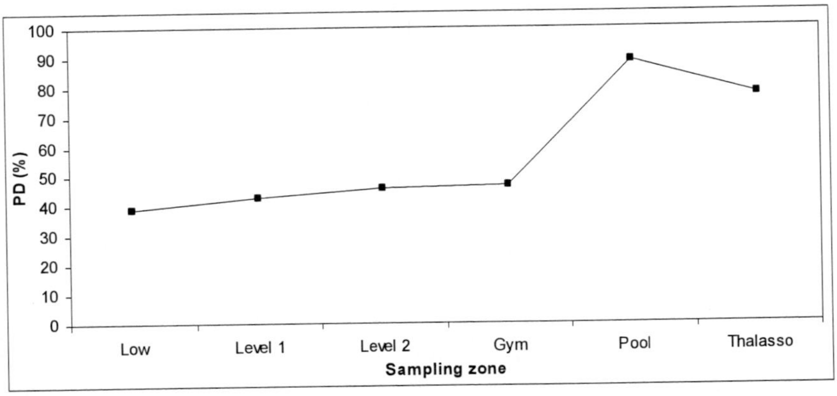

Figure 4. Percentage of dissatisfied persons at different zones of the spa

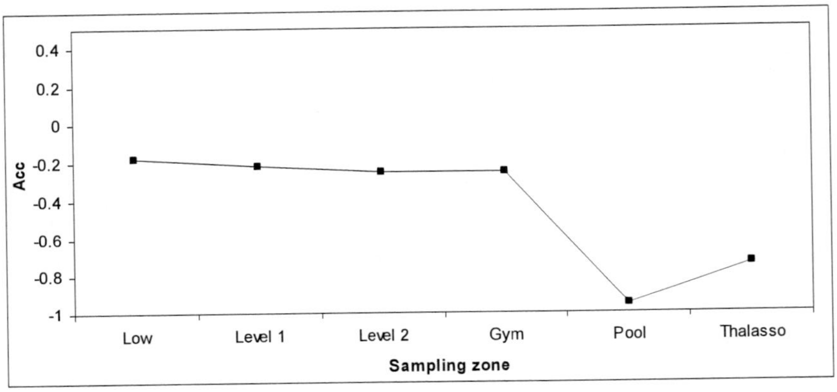

Figure 5. Air acceptability in the different zones of the spa

These values are reflected also by the conditions of perception of air humidity on mucous membranes (Acc). This index has shown poor indoor air perception in nearly all the zones, particularly in the pool. Despite this, in contrast to the expected results, in zones like Thalasso, a better acceptability was obtained than in the pools. This is probably because the discomfort caused by the relative humidity is compensated by the air temperature.

Finally, attention must be paid to the sudden temperature changes existing in each of the zones of high temperatures of costumes to the halls and corridor areas. Therefore, surveys were proposed to define the local comfort conditions present in each area.

3.3. Energy Saving

To conclude, the principal actuation ways that permit the reduction of energy consumption of such indoor ambiences, considering the temperature and the relative humidity, can be summarised. For example, to prevent energy loss, it is usually proposed to increase the relative humidity of the enclosures, thus reducing evaporation. Despite this, to avoid condensation on the outer walls, this solution is not always feasible given the possibility of moist air condensing if the relative humidity exceeds 50%, and this value is typically higher, in this humid Spanish region. Therefore, an increase in the indoor air temperature in the pool ambience alone is possible, but always within the established limit of 3 degrees higher than the water temperature. However, this is an extreme example of indoor conditions in Thalasso pools. As anticipated, such indoor ambiences will present a high percentage of dissatisfied and consequently, the residence time in these pools must be limited.

However, it is interesting to control the introduction of moist air into this indoor ambience and its energy implications. Initially, the renewal of indoor air proved to be a critical parameter for changes in energy consumption throughout the system. If renewal of the air were to be doubled, the power consumption would double. Next, the air velocity in the enclosure is to be considered and limited, as it is another parameter related to evaporation and therefore, with the energy lost in the installation.

Finally, other parameters like thermal and hygroscopic inertia of the buildings and pool water mass that also exert some influence on the overall energy consumption of the building. Presently, this influence is being defined with new software resources like TRNSYS [20] or HAM-tools [21, 22] which need to be validated with real sampled data to develop new optimal pool design and management.

4. Conclusions and Proposals of Actuation

After analysing the information from ASHRAE, ISO 7730, RITE and NTP 689, it is concluded that indoor air temperature depends on the pool water temperature, and that this in turn depends on the specific type of health effect that is to be achieved. From this perspective, the indoor conditions in the Thalasso room revealed a more extreme ambience with temperatures of 31.5°C and relative humidity of over 74.9%. These values imply a condensation risk on the outer walls and high energy consumption. However, the perception of air quality and indoor air acceptability showed a certain degree of local discomfort, therefore, low residence time will be required in such an ambience.

Finally, some energy saving procedures such as controlling the relative humidity of the pool ambience and employing pool barriers during the unoccupied period, are proposed. Future research works are warranted to control all these parameters based on the new HVAC control systems that control both temperature and relative humidity.

In these ambiences, studying the environmental quality with respect to energy saving is of great interest. The actual air changes and the presence of VOCs and CO_2 concentrations need to be measured and these values compared with RITE indications. However, the study of climatic data and its association with instant power consumption is also essential.

Another proposal of actions to really study indoor ambience is to conduct surveys that must be answered by occupants, workers and spa clients to get a real impression of the indoor conditions. Parameters like the thermal shock between costumes and dressing rooms or corridors and draft over the swimmers' wet body surface particularly, must be analysed in depth.

Acknowledgments

I would like to thank the University of A Coruña for sponsoring project 5230252906.541A.64902.

References

[1] Cimbiz, A; Bayazit, V; Hallaceli, H; Cavlak, U. The effect of combined therapy (spa and physical therapy) on pain in various chronic diseases. *Complementary therapies in Medicine*, 2005, 13, 244-250.

[2] Ledo, EMD. Mineral Water and Spas in Spain. *Clinics in Dermatology*, 1996, 14, 641-646.

[3] Routh, IB; Bhowmik, KR. Basic tenets of mineral water. A Glossary of Concepts relating to Balneology, Mineral Water, and the Spa. *Clinics in Dermatology*, 1996, 4, 6, 549-550.

[4] Soto, J; Fernández, PL; Quindós, LS; Gómez-Arozamena, J. Radioactivity in Spanish spas. *The Science of the Total Environment*, 1995, 162, 187-192.

[5] Lazzarín, RM; Longo, GA. Comparison of heat recovery systems in public indoor swimming pools. *Applied Thermal Engineering*, 1996, 16, 7, 561-570.

[6] Lam, JC; Chan, WW. Life cycle energy cost analysis of heat puma application for hotel swimming pools. *Energy Conversión and Management*, 2001, 42, 1299-1306.

[7] López González, LM; Miguez Tabarez, JL; Gándara Álvarez, M; Fernández Viar, P. Feasibility study for the installation of HVAC for a spa by jeans of energy recovery from thermal water-Part I. Analysis of conditions. *Renewable Energy*, 2001, 23, 123-134.

[8] Miguez Tabarez, JL; Gándara Álvarez, M; López González, LM; Fernández Viar, P. Feasibility study for the installation of HVAC for a spa by jeans of energy recovery from thermal water-Part I. Energy analysis. *Renewable Energy*, 2001, 23, 135-149.

[9] Mohammed Shah, M. Prediction of evaporation from occupied indoor swimming pool. *Energy and Buildings*, 2003, 35, 707-713.

[10] Asdrubali, F. A scale model to evaluate water evaporation from indoor swimming pools. *Energy and Buildings*, 2009, 41, 311-319.

[11] ISO, Thermal performance of buildings-calculation of energy use for space heating and cooling. ISO/DIS 13790:2005, *International Organization for Standardization*, Genea, 2005.

[12] ASHRAE, Energy Estimating and Modelling Methods, in: *ASHRAE Handbook, Fundamentals*, 2009, Atlanta.

[13] Normativa Técnica de Prevención de Riesgos. NTP 689. Piscinas de uso público (I). Riesgos y prevención. http:// www. diba. es/ salutpublica/ descarrega/ ntp_ 689% 20piscines.pdf

[14] Spanish Standard (RD 1027/2007), 2007. *Reglamento de Instalaciones Térmicas en los Edificios* (RITE).

[15] Standard SFS-EN 832. Thermal performance of buildings. Calculation of energy use for heating. *Residential buildings*. 1998.

[16] Standard SFS-EN ISO 13370, Thermal performance of buildings, heat transfer via the ground, *calculation methods*, 1998.

[17] Toftum, J; Jorgensen, AS; Fanger, PO. Upper limits for indoor air humidity to avoid uncomfortably humid skin. *Energy and Buildings*, 1998, 28, 1-13.

[18] Toftum, J; Jorgensen, AS; Fanger, PO. Upper limits of air humidity for preventing warm respiratory discomfort. *Energy and Buildings*, 1998, 28, 15-23.

[19] Ruiz, E; Martínez, PJ. Analysis of an open-air swimming pool solar heating system by using an experimentally validated TRNSYS model. *Solar Energy*, 2010, 84, 116-123.

[20] Orosa, JA. Energy Saving in HVAC of Spanish School Buildings. *Building and Environment*. Ed. Novascience. 2009 ISBN: 978-1-60876-128-9 https://www.novapublishers.com, 2009.

[21] Orosa, JA. New Strategies of HVAC System Design Based on Climatic Effect on Indoor Air Conditions in Buildings from Northwest of Spain. Energy and Buildings: Efficiency, *Air Quality, and Conservation. Ed. Novascience*. 2009. ISBN: 978-1-60741-049-2. https://www.novapublishers.com, 2009.

In: Relative Humidity: Sensors, Management…
Editor: John M. Castillo
ISBN: 978-1-61761-734-8

Chapter 10

MEASUREMENTS OF WATER UPTAKE PROPERTY OF ATMOSPHERIC AEROSOLS AT ELEVATED RELATIVE HUMIDITY

Kihong Park*[*] *and Jaesuk Kim
School of Environmental Science and Engineering, Gwangju Institute of Science and Technology (GIST), Gwangju, South Korea

ABSTRACT

The hygroscopic property (water uptake property at elevated relative humidity) of atmospheric aerosols is essential to understand the effect of aerosols on cloud formation by acting as cloud condensation nuclei (CCN) (i.e., hygroscopic particles grow into the CCN by absorbing water), which leads to climate change with cloud albedo effect. Also, hygroscopic property of particles is important to determine their transport behaviors and fates in the ambient atmosphere and to understnd their deposition pattern in the human respiratory system when they were inhaled. In this chapter, first, we describe a hygroscopicity tandem differential mobility analyzer (HTDMA) system in details to measure hygroscopic property of atmospheric aerosols in real time by measuring particle size change at an increased relative humidity. Then, we applied the HTDMA system to determine hygroscopic growth factors (i.e., ratio of particle size increase under high relative humidity) of various laboratory-generated particles of known composition and size to evaluate the performance of the HTDMA system. Finally, the hygroscopic GFs of particles in the ambient atmosphere at the relative humidity of 85% in various atmospheric environments (urban, rural, and coastal sampling sites) were measured and their results were compared. This will provide the estimation of chemical constituents of atmospheric aerosols in different environments, leading to a better understanding of their source characteristics.

---

* Corresponding author: Email: kpark@gist.ac.kr

Introduction

Atmospheric aerosols (particles suspended in the air) are of current interest due to their effects on climate change and human health (Dockery and Pope Iii, 1994). Particles in the air scatter or absorb incoming solar light affecting the earth's radiation balance (IPCC, 2007). They play an important role in cloud formation acting as cloud condensation nuclei. Also, they are involved in heterogeneous atmospheric chemistry and reaction. As a public concern, atmospheric particles have caused visibility impairment and can affect human health. When particles are inhaled by humans, they can cause respiratory and lung disease. Especially, fine (particles smaller than 2.5 um) and ultrafine (particles smaller than 100 nm) particles with high surface area-to-volume ratio (Oberdörster, 2000; Peters et al., 1997) can deeply penetrate into the human lung system, probably causing more toxic effect. The atmospheric particles are directly emitted by various sources such as vehicles, industry, biomass burning, sea spray, dusts, and etc (Andreae, 2007; Andreae, 2009). These directly emitted particles are considered as primary aerosols. Also, particles in the ambient atmosphere can be produced by a gas-to-particle conversion process. In this case, a variety of precursor gases experience a series of oxidation processes with available oxidizing agents (e.g., OH, O_3) in the ambient atmosphere, leading to production of low volatility and condensable products. The accumulation of that low volatility vapor leads to production of new particles (i.e., nucleation) or to their condensing on to preexisting particles at certain environmental conditions (relative humidity, T, etc.). The particles formed in this way are considered as secondary aerosols. Also, particles can be classified into natural aerosols (sea spray, volcano eruption, dust, and so on) and anthropogenic (man-made) aerosols (diesel exhaust, cooking, combustion, and so on), depending on their origin.

Aerosol particles have various physical and chemical properties depending on their source and formation/growth pathways. To find their source and to better understand the particle formation/growth mechanism and their effects on climate change and human health, their physical and chemical parameters (particle size, shape, density, mass, volume, number, hygroscopicity, volatility, chemical composition, etc.) must be determined (McMurry, 2000). Especially, the hygroscopic property (i.e., water uptake property at elevated relative humidity) of particles is important to understand the effect of aerosols on cloud formation by acting as cloud condensation nuclei (CCN) (i.e., particles are activated to cloud droplets by absorbing water under increased relative humidity) and to determine their transport behaviors and fates in the ambient atmosphere. For example, typically non-hygroscopic particles such as fresh soots produced by the combustion process do not absorb water, not acting as the CCN, while the sulfate particles, which are hygroscopic, can easily act as CCN. Also, the hygroscopic property of particles affects their deposition pattern in the human respiratory system when they are inhaled into the human body.

Various types of instruments have been developed to measure the hygroscopic property of atmospheric aerosols. The Environmental Scanning Electron Microscopic (ESEM) technique (Ebert et al., 2002) was used to measure the hygroscopic property of particles. In this technique, particles were first sampled onto substrates or grids. Then, the samples in the environmental cell were exposed to a humid condition at a regulated relative humidity, and the microscopy was used to look into the morphological change of particles with varying relative humidity. If particles do not experience any morphological change by absorbing water at increased relative humidity, these particles can be classified into non-hygroscopic particles, while if particles experience sudden change at the increased relative humidity, particles were classified into hygroscopic particles. However, in this technique, particles were exposed in a low vacuum condition, leading to evaporative sampling of artifacts, and it is off-line analysis (i.e., particles must be sampled first). The hygroscopicity Tandem Differential Mobility Analyzer (HTDMA) (McMurry et al., 1996; McMurry and Stolzenburg, 1989; Park et al., 2008a; Park et al., 2009a; Rader and McMurry, 1986) technique has been developed to determine the hygroscopic property of particles in real time. The detailed description of this technique was included in the experimental section. The HTDMA technique measures particle size change at varying relative humidity to determine the amount of water uptake in terms of particle size and volume (assuming spherical particles) in real time. To expose particles into different relative humidity environments, a relative humidity conditioner consisting of a saturator, mass flow controller and membranes was constructed. This technique also enables us to determine size-resolved hygroscopic property of particles and the mixing characteristics of complex aerosols.

In this study, we have developed the HTDMA system (Park et al., 2009a; Park et al., 2009b) which can cover a wider size range, including nanoparticles, than the conventional HTDMA system. A variety of aerosols of known size and composition were generated to evaluate the current HTDMA system. This also provides a hygroscopic database for various types of particles. Then, we applied this technique to measure the hygroscopic property of atmospheric particles under elevated relative humidity (85% relative humidity) in different environments such as urban and coastal sampling sites (Park et al., 2009b). This enables us to characterize various types of aerosols in terms of their water uptake property.

METHODS

The HTDMA system mainly consists of two differential mobility analyzers (DMA), a relative humidity conditioner, and condensation particle counter (CPC). Figure 1 shows a schematic of the current HTDMA system.

The DMA can classify particles according to their electrical mobility size. Particles were charged by a neutralizer before they entered the DMA. The neutralizer reduced the original charge level of particles and provided a known charge distribution for the particles. The DMA is a cylindrical tube with two flow inlets and outlets for aerosol flow and sheath (clean and filtered air) flow as shown in Figure 2 (Chen et al., 1998; Knutson and Whitby, 1975). The DMA has an inner electrode at the center where a certain voltage is applied with the outer electrode grounded. The charged particles in the aerosol flow will migrate into the center electrode with specific trajectory depending on the applied voltage and electrical mobility of particles. The electrical mobility of particles is related to their size. Thus, the DMA can classify particles according to their size.

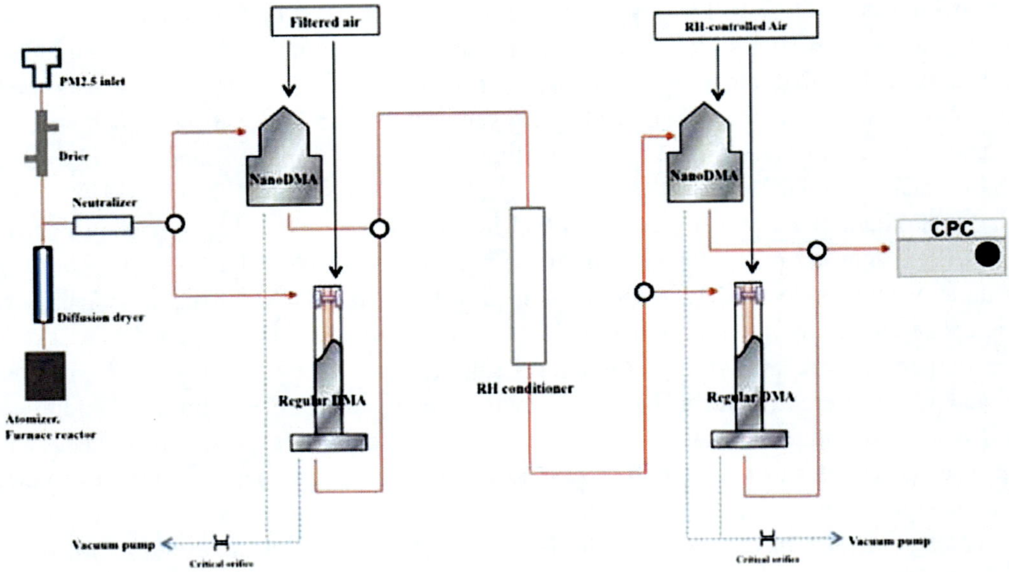

Figure 1. A schematic of the hygroscopicity tandem differential mobility analyzer (HTDMA) system

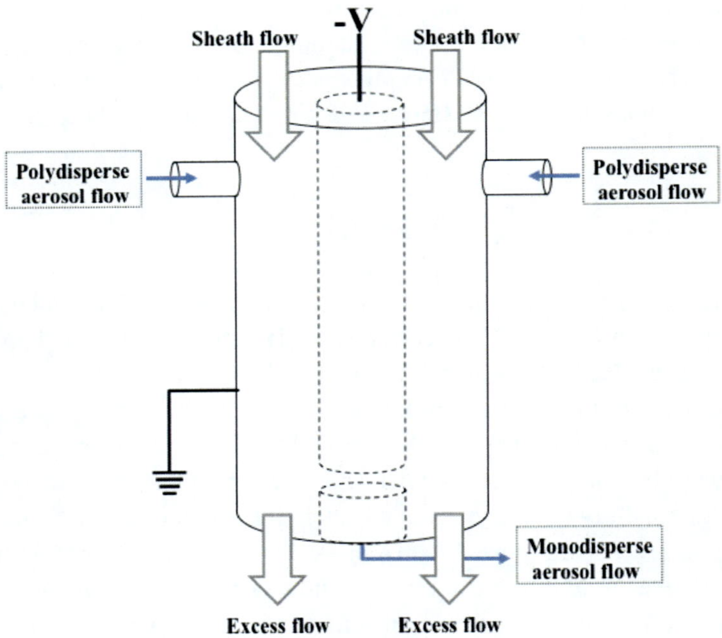

Figure 2. The schematic of the differential mobility analyzer (DMA)

The selected particles were sent into a relative humidity conditioner where particles experience various relative humidity conditions. To achieve a desired relative humidity condition, we controlled wet and dry air flow rates. The mixture of the dry and wet air was sent into the relative humidity conditioner. The relative humidity conditioner consists of porous membrane wall dividing the two streams. The aerosol flows through the center tube and the mixture of dry and wet air flows through the outside tube. Thus, the humid air diffuses into aerosol flow through the membrane. The schematic of the relative humidity conditioner is shown in Figure 3 (Sakurai et al., 2005). The wet air (relative humidity~100%) was produced by passing the compressed air through the saturator. The saturator was full of water in the sealed chamber where water liquid and vapor was equilibrated. The heated tape installed in the outside wall of the saturator chamber was used to maintain a fixed temperature (~30 °C). the mass flow controller was used to regulate the wet air flow rate. To control the relative humidity, mass flow rates of wet and dry airs was adjusted. In this way, particles of a certain size will experience a desired relative humidity condition when they passed through the relative humidity conditioner. The relative humidity sensor (GE sensing MMR31 and Omega HX94AV) and thermocouple was installed in the aerosol flow line downstream of the relative humidity conditioner to monitor and controal the relative humidity and temperature of the aerosol flow.

After particles experience the increased relative humidity condition, the physiochemical properties of particles will change. For example, hygroscopic NaCl particles will absorb water spontaneously at the specified relative humidity (75.3%) which is called as deliquescence relative humidity of NaCl particles (Tang and Munkelwitz, 1993). When this occurred, the NaCl particles will grow about 2 times. The particles after the relative humidity conditioner will enter the second DMA and CPC where particle size distribution was measured. The CPC provides particle number concentration by using the light scattering from particles (Agarwal and Sem, 1980). In the CPC, particles were saturated with butanol and grew into large particles in the condenser (i.e., supersaturated condition). These large particles scatter light easily that can be counted by laser and detector. By scanning the second DMA voltage, we are able to obtain particle number concentration for the given particle size, leading to provide particle number size distribution. Thus, by comparing the size of particles selected at the first DMA and that obtained from the second DMA and CPC, we are able to determine particle size change as a function of the relative humidity. One example for NaCl particles measured by our HTDMA system was shown in Figure 4. We selected 50 nm NaCl particles and exposed them to various relative humidity conditions. The resultant particle size was measured with the second DMA and CPC. The ratio of final size at the increased relative humidity to the initial size at the dry condition can be defined by particle hygroscopic growth factor (HGF). Figure 4 shows the change of particle size distribution as a function of the relative humidity for 50 nm NaCl particles. The NaCl particles suddenly absorbed water at the 76% relative humidity (i.e., particle size distribution moved to left as seen in Figure 4), which is close to the theoretical deliquescence relative humidity of NaCl. Below the 76% relative humidity, there is no change of particle size distribution (i.e., particles do not aborb water at those relativie humidities.)

To produce various particles in the laboratory, atomization and condensation/evaporation methods were used. An atomizer was used to generate particles from solution. A known amount of solution (e.g., 0.1 wt % NaCl in water) was atomized into water droplets by using the compressed air and these droplets passed through a series of diffusion driers where solvents are evaporated and crytallized to solid particles in the air. To generate nanoparticles less than 30 nm, the furnace reactor was used. The sample in the boat was placed in the furnace under high temperature. The sample will be evaporated and the vapor cool down by mixing with cool air leading to produce very small nanoparticles. Particles produced in this way were sent directly into the HTDMA system.

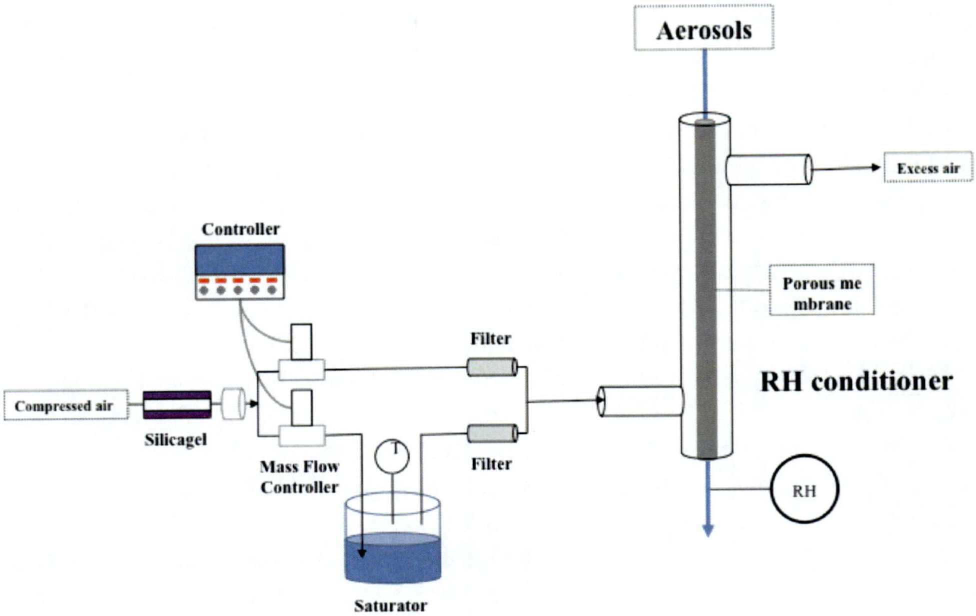

Figure 3. A schematic of the relative humidity-controlled system

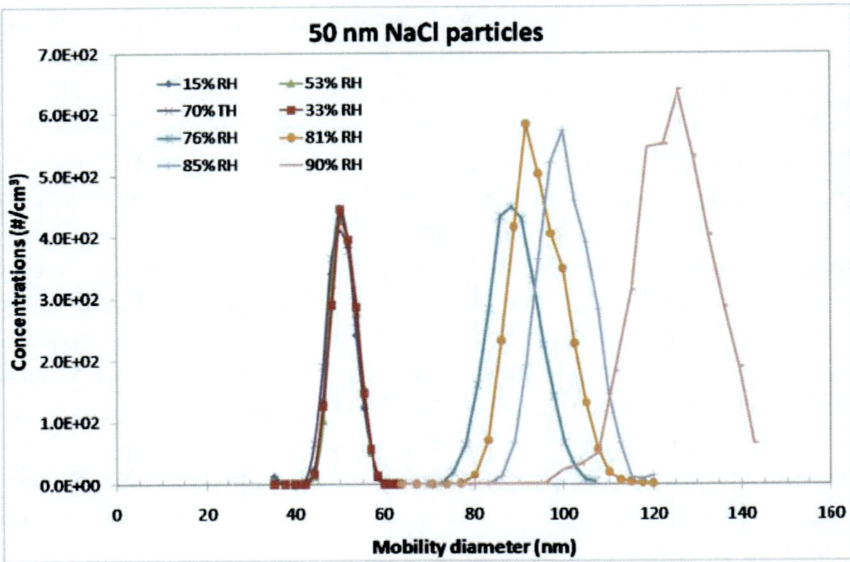

Figure 4. Particle size distribtuion change as a function of the relative humidity for 50 nm NaCl particles

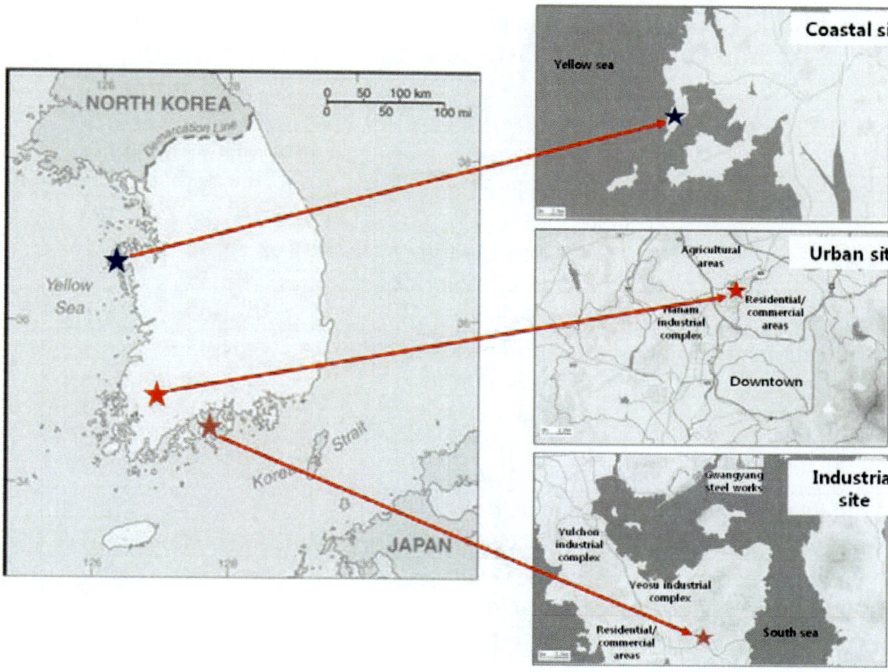

Figure 5. A map of several sampling sites for hygroscopicity measurement of atmopsheric aerosols

For atmospheric aerosol measurements, urban and coastal sites in Korea were selected. The map of the sampling sites including possible local sources was shown in Figure 5. The atmospheric aerosols was introduced through PM2.5 inlet where particles larger than 2.5 um were removed. Then, particles were dried out and passed through the neutralizer where particles were charged to a known charge distribution. The charged particles continuously entered the HTDMA system. At a given relative humidity (85% relative humidity), it took about 2 min to measure the HGF of particles of a certain size. In addition, to look into morphology and elemental composition of particles, we collected particles on to TEM grids and then these particles were examined by the transmission electron microscopy/energy dispersive X-ray spectroscopy (TEM/EDS).

Results and Discussion

We applied the HTDMA system described above to measure the hygroscopic growth factor (HGF at the relative humidity of 85%) of various laboratory-generated particles that can be candidate species in the atmospheric aerosols. Depending on chemical properties of particles, various HGFs were obtained. Table 1 summarizes the HGF of particles tested here. We classify those particles into four groups (high-hygroscopic, moderate-hygroscopic, low-hygroscopic, and non-hygroscopic).

Table 1. The hygroscopic growth factor (HGF) of various laboratory-generated particles at 85% relative humidity

Hygroscopicity	Chemical species	Hygroscopic growth factor at 85% relative humidity*
High-hygroscopic	Sodium chloride	2.00
	Sulfuric acid	1.80
	Potassium chloride	1.68
	Magnesium chloride	1.57
	Humic acid	1.69
Moderate-hygroscopic	Ammonium sulfate	1.48
	Calcium chloride	1.42
	Ammonium nitrate	1.34
	Malonic acid	1.35
	Levoglucosan	1.25
Low-hygroscopic	Iodine oxides	1.05
	Glutaric acid	1.04
Non-hygroscopic	Succinic acid	1.00
	Oxalic acid	1.00
	N-alkanes	1.00
	Soot	1.00

* Hygroscopic growth factors of 50 nm particles

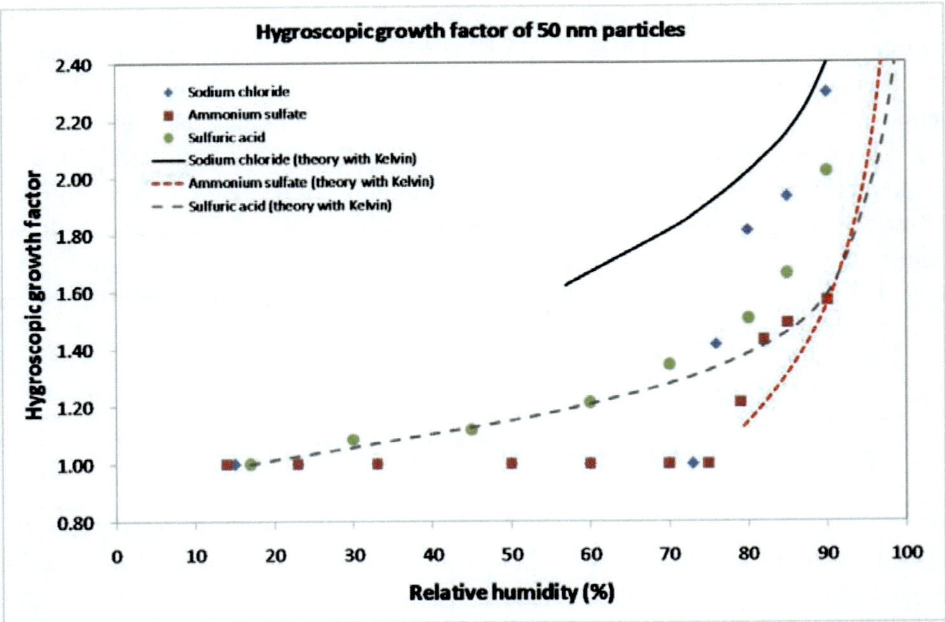

Figure 6. Comparison of the measured HGF with the HTDMA technique with theoretical values for 50 nm sodium chloride, ammonium sulfate, and sulfuric acid particles

Elemental carbon or black carbon particles typically produced by combustion process are non-hygroscopic. Some volatile organics such as n-alkanes are also non-hygroscopic. Dicarboxlyic acids are found to be moderate-hygroscopic or low hygroscopic. Sulfate, sulfuric acid, and salt particles are high-hygroscopic. Hygroscopic property of some inorganic particles are theoretically known. The deliquescence relative humidity (deliequescence relative humidity) and growth factor of such particles as a function of relative humidity can be theoretically calculated. We compared these values with our measured data with the HTDMA technique for 50 nm sodium chloride, ammonium sulfate, sulfuric acid particles as shown in Figure 6 (Biskos et al., 2006; Hämeri et al., 2001). The measured deliequescence relative humidity agreed well within 1% and the GF were in a good agreement within 12%. In case of sulfuric acid particles there was no deliequescence behavior like sodium chloide and ammonium sulfate particles (i.e., the sulfuric acid particles grow gradualy as the relative humidity increases). Our data suggest that the HTDMA technique can determine the hygroscopic property of atmospheric particles with such accuracy.

The HTDMA technique was then applied to determine hygroscopic property of atmospheric ultrafine particles. The size and number of particles were continuously monitored with the DMA and CPC. When ultrafine particle concentration increased, the HTDMA system was triggered to measure hygroscopic growth factor of those particles (i.e., we selected ultrafine particles of mode diameter by the first DMA). This enables us to estimate chemical species of ultrafine particles. Figure 7 shows hourly variation of particle size distribution on November 10, 2008. As shown, ultrafine particle concentration rapidly increased around 11 am. A strong photochemical activity in the afternoon has been known to contribute to form ultrafine particles. When those ultrafine particles are formed, the HTDMA system was operated to measure their hygroscopic property. At this time, the hygroscopic growth factor was found to be 1.50±0.03. Before this event occurred, the hygroscopic growth factor of 50 nm particles was 1.24±0.01. The increase of the hygroscopic growth factor suggests that the ultrafine particles contain a significant amount of sulfates (i.e., sulfates contributed to the formation and growth of ultrafine particles at this time).

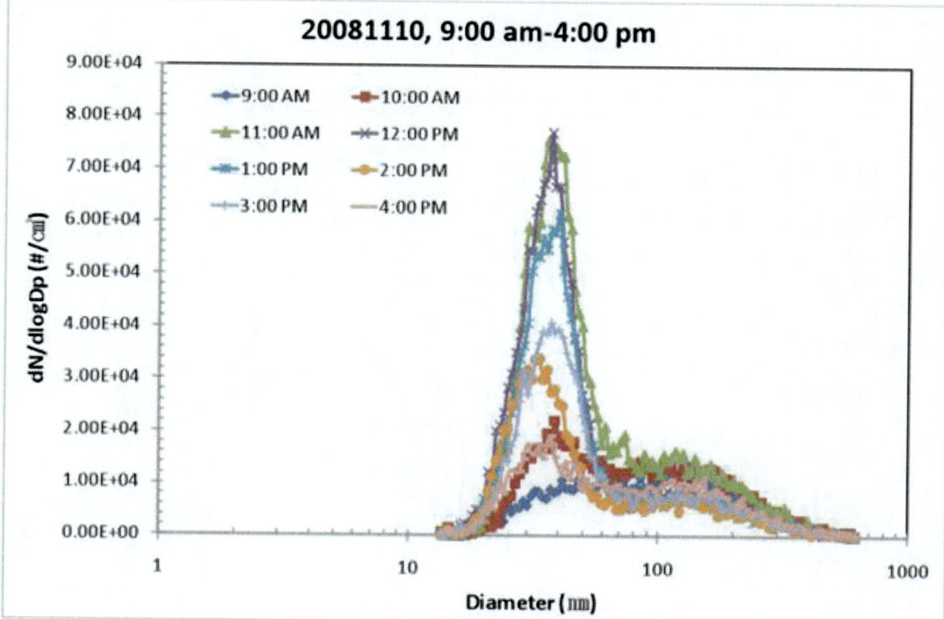

Figure 7. Houtly variation of particle size distribtuion on 11/10/2008 at the urban Gwangju site

Similar ultrafine particle events (i.e., photochemical event) in other days are shown in Figure 8. The x-axis is the time of day, the y axis is the particle size, and the color indicates particle concentration (dN/dlogDp). As it can be seen, the significant increase of ultrafine particles occurred in the afternoon when the solar radiation was strong and ozone concentration was high, which is the typical case of photochemically-induced ultrafine particles (Park et al., 2009b; Park et al., 2008b).

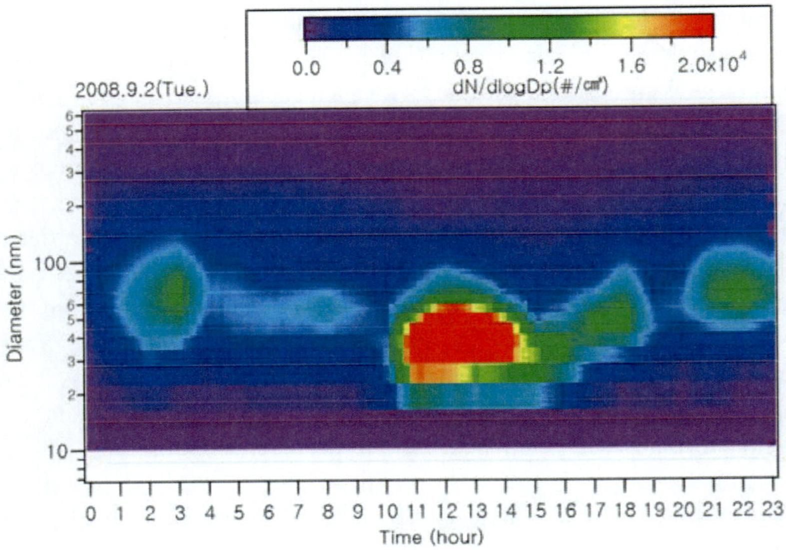

Figure 8. Diurnal variations of particle size distribution with diameter on the y-axis, time of day on the x-axis, and particle concentration (dN/dlogDp) by shading contour in photochemical event at the urban Gwangju site

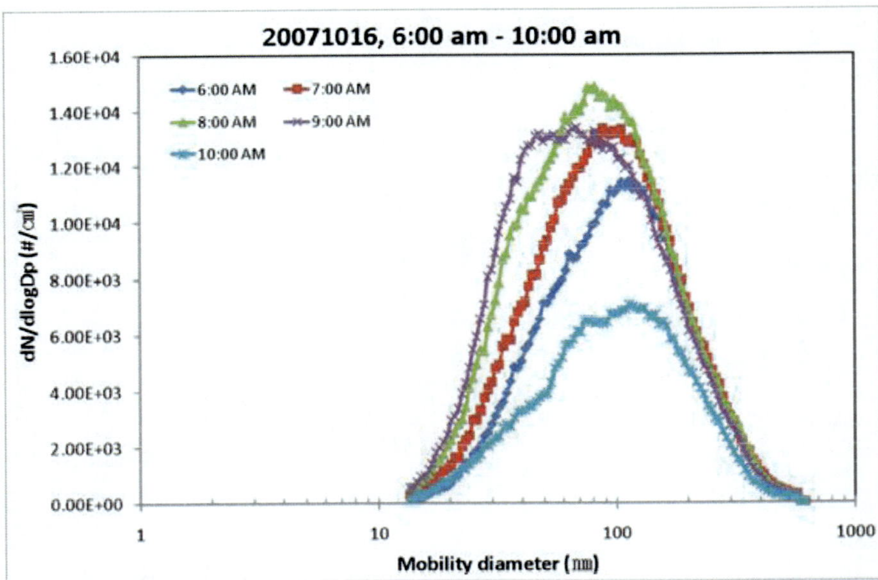

Figure 9. Hourly variation of particle size distribution during "combustion event" on October 16, 2007 at the urban Gwangju site

Figure 9 shows that the increase of ultrafine particles occurred at ~8 am. The time for increase of particles coincided with morning traffic hours and the time for the increase of CO concentration that is a tracer for combustion event. Their hygroscopic growth factor was 1.06±0.02, which is pretty low. This suggests that these particles mainly contain non-hygroscopic elemental carbon or black carbon. We can refer to this event as combustion event.

Similar combustion events observed during the sampling period are shown in Figure 10. In addition to morning traffic hours, this event also occurred in the evening time. The increase of CO in the event time might be caused by combined sources from residential heating, cooking, and traffics. Also, The decrease of the mixing layer in the evening might cause stagnant air condition leading to high concentration of particles.

We summarized the measured hygroscopic factors of ultrafine particles when ultrafine particle concentration increased at the urban site during the sampling period in Table 2. Also, peak size and particle number concentration were included. The measured hygroscopic factors of ultrafine particles formed in the afternoon under a strong photochemical activity were smaller then those of pure sulfate particles, suggesting that these particles were internally mixed with less hygroscopic species.

Figure 11 compares the hygroscopic growth factors of ultrafine particles observed in the afternoon at the urban and coastal sites. Data showed that the hygroscopic growth factor of ultrafine particles observed at the coastal site was much higher than that at the urban site. This suggests that ultrafine particles observed at the coastal site were less mixed with other less hygroscopic species. Also, hygroscopic salts originated from ocean may be included in those particles leading to be more hygroscopic (O'Dowd and De Leeuw, 2007; O'Dowd et al., 2002).

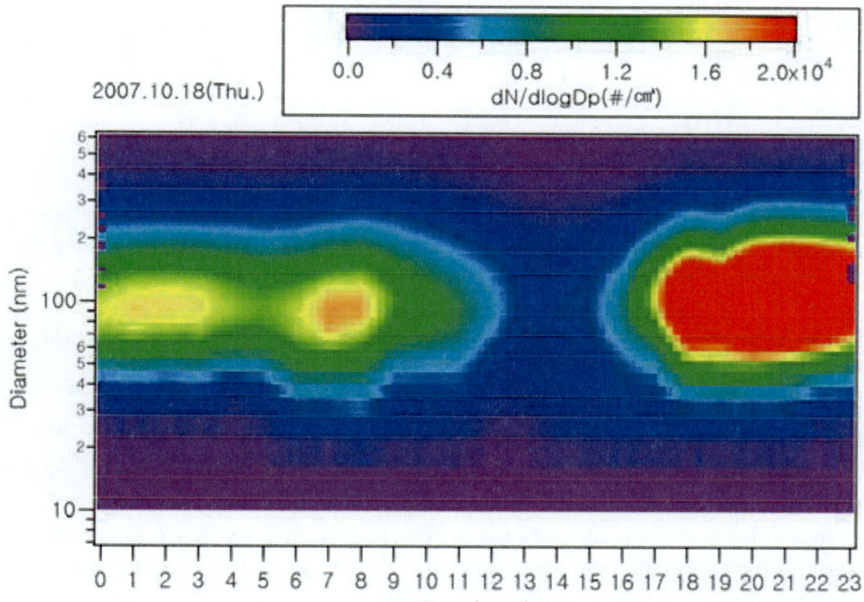

Figure 10. Diurnal variations of particle size distribution with diameter on the y-axis, time of day on the x-axis, and particle concentration (dN/dlogDp) by shading contour in the combustion event at the urban Gwangju site

Table 2. Summary for the time, HGF, and particle number concentration for ultrafine particles events observed at the urban Gwangju site

Date	Local time	Size (Centre)	HGF	N (13~100 nm)	N (13~600 nm)
10/14/2007	13:00 ~ 18:00	25	1.05±0.02	4.70E+03	7.55E+03
		30	1.05±0.02		
10/15/2007	15:00 ~ 18:00	40	1.07±0.02	5.61E+03	9.17E+03
10/16/2007	13:00 ~ 19:00	20	1.07±0.03	4.39E+03	7.44E+03
05/05/2008	08:00 ~ 15:00	50	1.33±0.04	6.76E+03	8.87E+03
05/06/2008	13:00 ~ 17:00	25	1.24±0.04	7.02E+03	9.40E+03
05/08/2008	13:00 ~ 18:00	35	1.22±0.02	6.47E+03	1.05E+04
05/11/2008	12:00 ~ 18:00	40	1.21±0.02	7.41E+03	9.16E+03
05/15/2008	13:00 ~ 18:00	35	1.17±0.02	5.92E+03	7.60E+03
05/17/2008	12:00 ~ 17:00	20	1.27±0.03	6.29E+03	9.32E+03
		30	1.15±0.06		

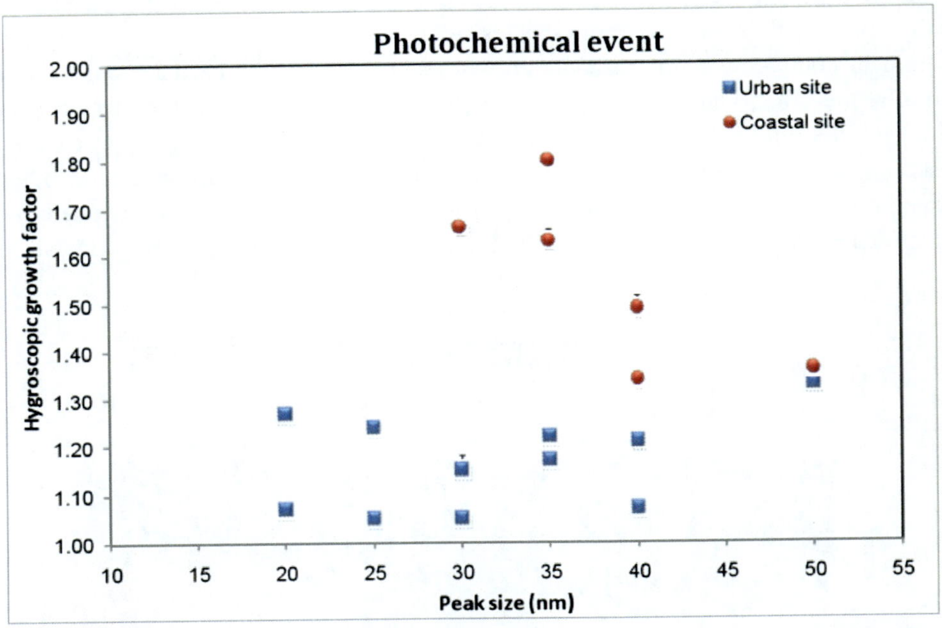

Figure 11. Comparison of hygroscopic growth factors in photochemical event between urban and coastal sites.

Morphological and elemental property of these particles were examined by the TEM/EDS. Several different types of particles in terms of morphology and elemental composition were observed. We classify these particles into several major types: spherical sulfate particles, spherical organic carbon particles, agglomerated soot particles, and non-spherical metal oxide particles (Tumolva et al., 2010). Figure 12 shows these types of particles with their elemental composition obtained from the EDS.

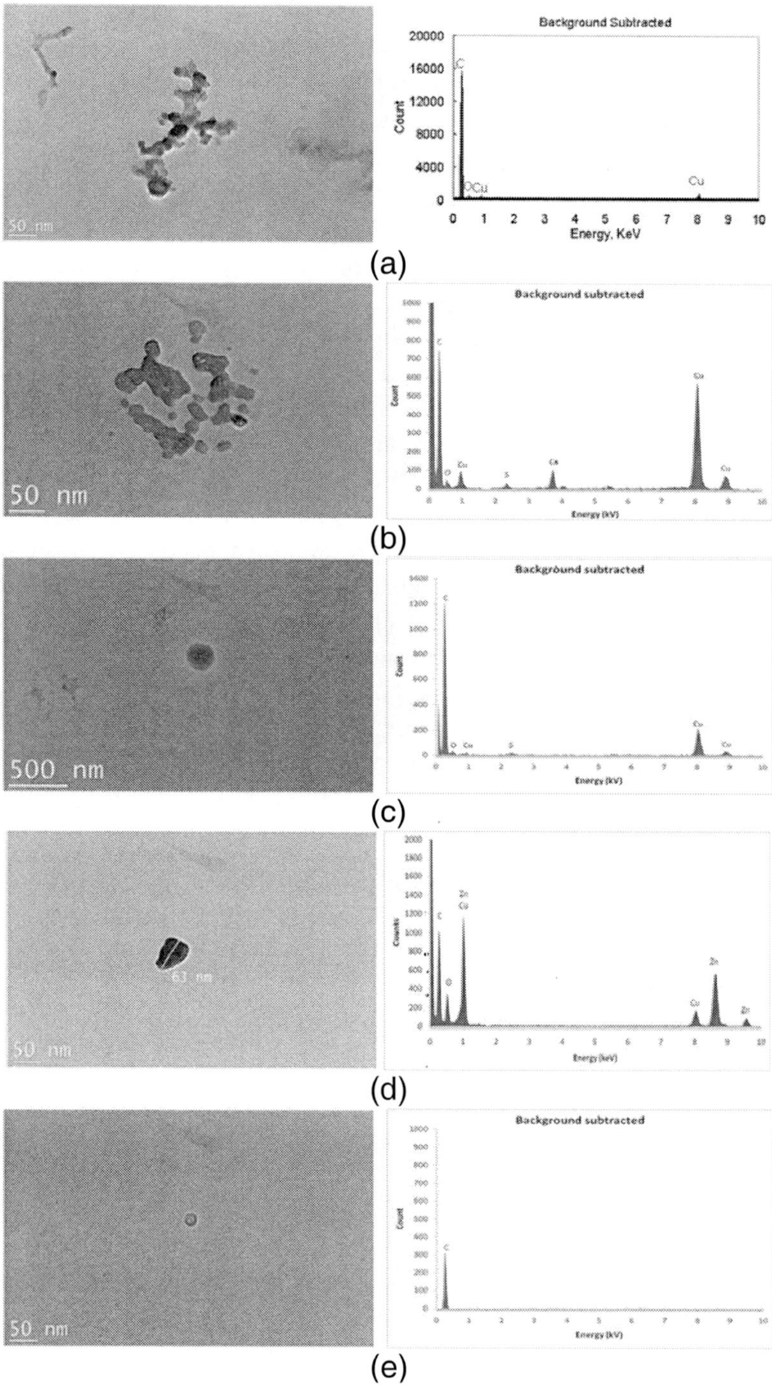

Figure 12. Morphology and elemental composition of particles: (a) freshly agglomerated soot particle, (b) aged agglomerated soot particle, (c) spherical sulfate particle, (d) non-spherical metal particle, and (e) spherical organic carbon particle.

CONCLUSION

The hygroscopic tandem differential mobility analyzer (HTDMA) system to determine the hygroscopic property of atmospheric aerosols in real time by measuring particle size change at an increased relative humidity was described in detail. The relative humidity conditioner system in the HTDMA technique had particles exposed to various relative humidity environments, enabling us to look into the water uptake property of particles at increased relative humidities. Hygroscopic growth factors (HGF) (i.e., ratio of particle size increase under high relative humidity) of various laboratory-generated particles of known composition and size were compared with theoretical values, leading to the evaluation of the performance of the HTDMA system. The measured deliquescence relative humidity agreed well within 1% and the GF were in a good agreement within 12%. The HTDMA technique was then applied to determine the hygroscopic property of atmospheric ultrafine particles. The size and number of particles were continuously monitored with the DMA and CPC. When ultrafine particle concentration increased, the HTDMA system was used to measure hygroscopic growth factor of those particles. The significant increase of ultrafine particles occurred in the afternoon when the solar radiation was strong and ozone concentration was high (i.e., the typical case of photochemically-induced ultrafine particles). The HGF of ultrafine particles in the photochemical event was significanty higher than those in the combustion event when the time for increase of particles coincided with morning traffic hours and the time for the increase of CO concentration. In the combustion event, the HGF was found to be 1.06±0.02, suggesting that these ultrafine particles mainly contained non-hygroscopic elemental carbon or black carbon.

ACKNOWLEDGMENT

The review with new data analysis was supported by the National Research Foundation of Korea Grant funded by the Korean Government (NRF-2010-013).

REFERENCES

Agarwal, J. K. & Sem, G. J. (1980). Continuous flow, single-particle-counting condensation nucleus counter. *Journal of Aerosol Science*, *11(4)*, 343-357.

Andreae, M. O. (2007). Aerosols before pollution. *Science*, *315(5808)*, 50-51.

Andreae, M. O. (2009). A new look at aging aerosols. *Science*, *326(5959)*, 1493-1494.

Biskos, G., Paulsen, D., Russell, L. M., Buseck, P. R. & Martin, S. T. (2006). Prompt deliquescence and efflorescence of aerosol nanoparticles. *Atmospheric Chemistry and Physics*, *6(12)*, 4633-4642.

Centre, G. P. E. P. M. (2007). *Pearl River Delta Regional Air Quality Monitoring Network-A Report of Monitoring Results in*, 2006.

Chen, D. R., Pui, D. Y. H., Hummes, D., Fissan, H., Quant, F. R. & Sem, G. J. (1998). Design and evaluation of a nanometer aerosol differential mobility analyzer (Nano-DMA). *Journal of Aerosol Science*, *29(5-6)*, 497-509.

Dockery, D. W. & Pope Iii, C. A. (1994). Acute respiratory effects of particulate air pollution. *Annual Review of Public Health, 15*, 107-132.

Ebert, M., Inerle-Hof, M. & Weinbruch, S. (2002). Environmental scanning electron microscopy as a new technique to determine the hygroscopic behaviour of individual aerosol particles. *Atmospheric Environment, 36(39-40)*, 5909-5916.

Hämeri, K., Laaksonen, A., Väkeva, M. & Suni, T. (2001). Hygroscopic growth of ultrafine sodium chloride particles. *Journal of Geophysical Research D*: Atmospheres, *106(D18)*, 20749-20757.

IPCC, (2007). Climate Change. IPCC *Fourth Assessment Report*. Cambridge University Press, New York.

Knutson, E. O. & Whitby, K. T. (1975). Aerosol classification by electric mobility: apparatus, theory, and applications. *Journal of Aerosol Science, 6(6)*, 443-451.

McMurry, P. H. (2000). A review of atmospheric aerosol measurements. *Atmospheric Environment, 34(12-14)*, 1959-1999.

McMurry, P. H., Litchy, M., Huang, P. F., Cai, X., Turpin, B. J., Dick, W. D. & Hanson, A. (1996). Elemental composition and morphology of individual particles separated by size and hygroscopicity with the TDMA. *Atmospheric Environment, 30(1)*, 101-108.

McMurry, P. H. & Stolzenburg, M. R. (1989). On the sensitivity of particle size to relative humidity for Los Angeles aerosols. *Atmospheric Environment, 23(2)*, 497-507.

O'Dowd, C. D. & De Leeuw, G. (2007). Marine aerosol production: A review of the current knowledge. Philosophical Transactions of the Royal Society A: Mathematical, *Physical and Engineering Sciences, 365(1856)*, 1753-1774.

O'Dowd, C. D., Jimenez, J. L., Bahreini, R., Flagan, R. C., Seinfeld, J. H., H채merl, K., Pirjola, L., Kulmala, M., Jennings, S. G. & Hoffmann, T. (2002). Marine aerosol formation from biogenic iodine emissions. *Nature, 417(6889)*, 632-636.

Oberdörster, G. (2000). Toxicology of ultrafine particles: In vivo studies. Philosophical Transactions of the Royal Society A: Mathematical, *Physical and Engineering Sciences, 358(1775)*, 2719-2740.

Park, K., Dutcher, D., Emery, M., Pagels, J., Sakurai, H., Scheckman, J., Qian, S., Stolzenburg, M. R., Wang, X., Yang, J. & McMurry, P. H. (2008a). Tandem measurements of aerosol properties - A review of mobility techniques with extensions. *Aerosol Science and Technology, 42(10)*, 801-816.

Park, K., Kim, J. S. & Miller, A. L. (2009a). A study on effects of size and structure on hygroscopicity of nanoparticles using a tandem differential mobility analyzer and TEM. *Journal of Nanoparticle Research, 11(1)*, 175-183.

Park, K., Kim, J. S. & Park, S. H. (2009b). Measurements of hygroscopicity and volatility of atmospheric ultrafine particles during ultrafine particle formation events at urban, industrial, and coastal sites. *Environmental Science and Technology, 43(17)*, 6710-6716.

Park, K., Park, J. Y., Kwak, J. H., Cho, G. N. & Kim, J. S. (2008b). Seasonal and diurnal variations of ultrafine particle concentration in urban Gwangju, Korea: Observation of ultrafine particle events. *Atmospheric Environment, 42(4)*, 788-799.

Peters, A., Wichmann, H. E., Tuch, T., Heinrich, J. & Heyder, J. (1997). Respiratory effects are associated with the number of ultrafine particles. *American Journal of Respiratory and Critical Care Medicine, 155(4)*, 1376-1383.

Rader, D. J. & McMurry, P. H. (1986). Application of the tandem differential mobility analyzer to studies of droplet growth or evaporation. *Journal of Aerosol Science, 17(5)*, 771-787.

Sakurai, H., Fink, M. A., McMurry, P. H., Mauldin, L., Moore, K. F., Smith, J. N. & Eisele, F. L. (2005). Hygroscopicity and volatility of 4-10 nm particles during summertime atmospheric nucleation events in urban Atlanta. *Journal of Geophysical Research D, Atmospheres, 110(22)*, 1-10.

Tang, I. N. & Munkelwitz, H. R. (1993). Composition and temperature dependence of the deliquescence properties of hygroscopic aerosols. *Atmospheric Environment - Part A General Topics, 27 A(4)*, 467-473.

Tumolva, L., Park, J. Y., Kim, J. S., Miller, A. L., Chow, J. C., Watson, J. G. & Park, K. (2010). Morphological and elemental classification of freshly emitted soot particles and atmospheric ultrafine particles using the TEM/EDS. *Aerosol Science and Technology, 44(3)*, 202-215.

In: Relative Humidity: Sensors, Management…
Editor: John M. Castillo

ISBN: 978-1-61761-734-8

Chapter 11

MOISTURE BUFFERING CAPACITY OF HYGROSCOPIC MATERIALS

Marco D'orazio, Simona Cerolini and Alessandro Stazi
Department of Architecture, Construction and Structures (DACS),
Faculty of Engineering, Polytechnic University of Marche, Ancona, Italy

ABSTRACT

The demand for high performance buildings, with regard to thermal comfort and energetic efficiency, is causing a progressive reduction of the envelope's air permeability, the increase of indoor RH% in occupied buildings and the appearance of problems linked with high humidity levels (deterioration of building materials, mould growth and biological organisms proliferation, respiratory discomfort, skin humidity and unacceptable perceived air quality).

This research investigates the possibility to use highly absorbing materials to dampen indoor RH% variations. The practical MBV of sodium polyacrilate, cellulose-based material, perlite and gypsum is evaluated for a daily cyclic exposure that alternates high (75%) and low (33%) RH% levels for 8h and 16h respectively. The adjustment velocity to RH% variations and the presence of hysteretic phenomena are also presented. The cellulose-based material proves to be the most suitable for moisture buffering applications. Starting from this material's properties, the effect of thickness, vapour resistance factor (μ) and mass surface exchange coefficient (Z_v) on sorption capacity is evaluated by the use of a numerical model.

1. INTRODUCTION

Indoor humidity is mainly affected by the presence of sources of water (all human activities give off some vapour), by the ventilation and infiltration rate through the envelope and by the vapour adsorption capacity of the materials of the room. New buildings are still constructed with little porous and water absorbent materials and coated with almost impermeable paints.

Furthermore, the recent demand for controlling energetic consumptions and gas emissions has contributed to the progressive increase of air tightness of buildings: in order to improve the thermal behaviour of the building envelope, thick layers of insulating material are introduced in walls and roofs even in mild climates and not very air permeable windows are used. These conditions as a whole reduce the possibility of vapour exchange between the indoor air and the external environment through air leakages and cause a rise in indoor relative humidity levels, which negatively affects building materials' durability, human comfort and health.

The control of humidity conditions is particularly important to avoid the risks of toxic micro-organisms and moulds growth on the interior surfaces of building components, but the introduction of HVAC systems providing an adequate mechanical ventilation doesn't seem to be a solution, particularly in small, low-density interiors, such as houses, where it would be a source of noise for people living there and a cause of a further rise in energetic consumptions.

A promising strategy to meet with the increasing vapour load comes from the sector of cultural heritage preservation, and consists in the use of materials with hygroscopic sorption capacity in order to dampen humidity variations.

In this research, the dynamic behaviour of two building materials (gypsum and perlite), exposed to cyclic step-changes in the relative humidity, is compared with the performance of highly hygroscopic materials used in the industrial sector (cellulose based material and sodium polyacrilate), in order to verify the ability of materials already used in interior surface finishes to affect indoor humidity conditions and to consider the possibility to reduce the moisture buffering exposed area, by selecting highly performing materials. The comparison between the different materials is based on their *Moisture Buffer Value (MBV)* ($MBV_{practical}$ [kg/(m^2·% RH)]), defined as "the amount of water that is transported in or out of a material per open surface area, during a certain period of time, when it is subjected to variations in relative humidity of the surrounding air" [Rode, 2005].

The possibility to use the moisture buffering materials in a real building is then linked with the ability of such materials to maintain their sorption properties. The trend of the variations in weight of the specimens at the end of each cycle, shows the presence of any hysteretic phenomena in the tested materials and is presented in this chapter.

Finally, the effect of thickness, of vapour resistance factor μ [-] and of mass surface exchange coefficient Z_v [s/m] on the maximum and final water content of a system exposed to indoor RH% fluctuations is investigated by the use of a numerical model, io order to improve the system's efficiency.

2. Materials and Methodologies

2.1. Materials

In this study the MBV of two materials used in interior surface finishes and of two industrial products with high sorption capacity is determined.

In the first group gypsum and perlite are considered.

The gypsum specimen is a perforated panel used as false ceiling with sound proofing features. The expanded perlite is commonly added to walls, floors and roofs' interior finishes to increase their heat insulation.

In the group of the industrial materials sodium polyacrilate and a cellulose based material are considered.

Sodium polyacrilate is a super adsorbent polymer able to adsorb water up to 100 times its weight. It is currently used in several industrial sectors ranging from personal hygiene, medical field and cosmetics to the packaging of perishables and nursery-gardening.

Cellulose-based material is a combination of cellulose and superadsorbent polymers, yet used in personal hygiene field.

2.2. Experimental Facility

The moisture buffer performance of the specimens is evaluated according to the method for practical categorization of materials based on *MBV*, developed in the NORDTEST Project [Rode, 2005].

For this test, the materials, except for the gypsum specimen, are wound in a nonwoven fabric and then placed in a non absorbing plastic container to facilitate weighing procedure. The container is closed above with a perforated sheet metal to simulate the presence of a surface closure and sealed with aluminium adhesive tape. The edges and the back-sides of the gypsum specimen are sealed with a sheet of polyethylene and aluminium adhesive tape in order to obtain one-dimensional moisture flow (Figure 1, 2).

An empty container is also tested to assure the negligible hygroscopicity of the closer and sealing systems.

The samples are pre-conditioned at 23±0,3°C and 50±3 %RH in the climatic chamber Angelantoni CH250 and the weight of the specimens at equilibrium is determined.

Table 1 summarizes the exposure area and the weight at equilibrium for each specimen.

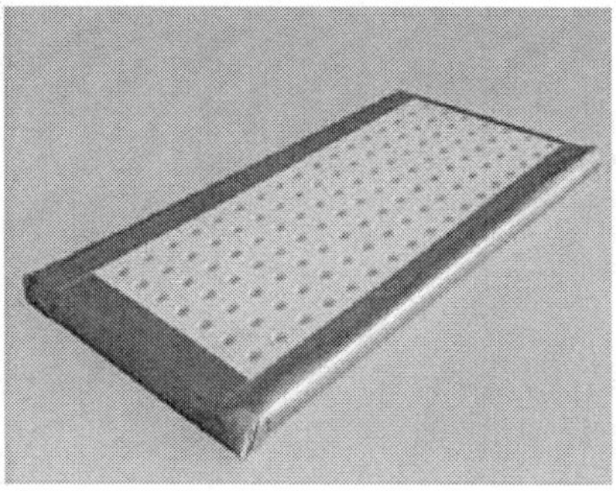

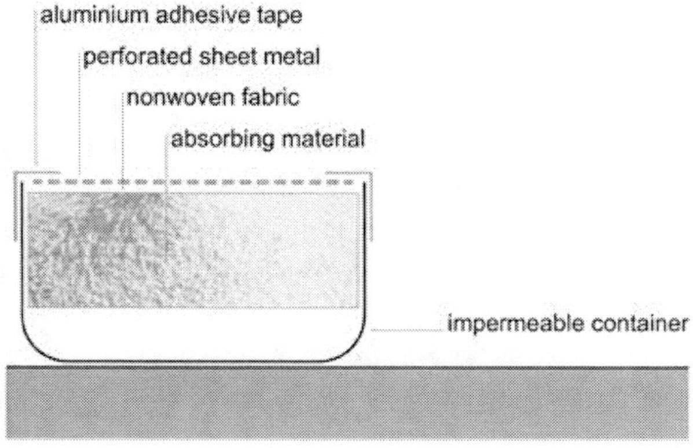

Figure 1, 2. Pictures showing the closing and sealing systems used for gypsum board (1) and for the other samples (2)

Table 1. Specimen's exposure area and weight at equilibrium after the pre-conditioning at 23±0,3°C and 50±3 % RH in climatic chamber

Material	Exposure area [m^2]	Weight at equilibrium [g]
Perlite	0,025	369,25
Cellulose	0,025	69,60
Gypsum	0,052	368,70
Sodium Polyacrilate	0,025	2041,30

According to the NORDEST method [Rode, 2005], the materials are exposed to cyclic step-changes that alternate high levels (75% for 8h) and low levels (33% for 16h) of relative humidity. The duration of the entire cycle is 24 hours, that is a daily exposure.

During the test the temperature is held constant at 23±0,3 °C. The amount of water absorbed and released by the materials for each step is determined monitoring the change in weight of the specimen and the average mass-change between the absorption and desorption phases is calculated for each cycle. The practical MBV is the mean mass-change value of the last 3 cycles, referred to the specimen's exposure area and to the RH-variation (42% RH).

2.3. Numerical Investigations

The selection of the tested materials is carried out depending on the exposed criteria and the characteristics of the specimens are determined according to the NORDTEST Project [Rode, 2005]. Actually, there is a great variety of porous materials and different exposure conditions, that may not be all reproduced by means of experimental activities.

Therefore, a numerical model may be used to quickly investigate the effect of material's thickness, vapour resistance factor μ [-] and surface exchange coefficient Zv [s/m] on the moisture buffering capacity of a material, enabling to define an optimum functioning range for these quantities and to detect the classes of materials that are more suitable to dampen RH% variations.

For this purpose the software *Delphin4* is used. This program, developed by the Institute of Building Climatology of the Technical University of Dresden [Grunewald, 2000], is able to model the coupled heat, air, salt and moisture transport that occurs in porous building materials, allowing to investigate the thermal and hygric behaviour of constructive building details for arbitrary standard and natural climatic boundary conditions.

In this case the problem is described by a single layer of homogeneous material exposed to a one-dimensional flux. The material has a surface area of $1m^2$ and a variable thickness (Table 2) discretisated with smaller volume elements at the boundaries.

In order to show the dependence of the moisture buffering capacity on the panel's thickness, this quantity is varied from 20 mm to 140 mm and the maximum and final (at the end of the simulation period) water content stored up in the material are considered.

The vapour resistance factor μ [-] and the surface exchange coefficient Zv [s/m] are varied separately starting from the characteristics of an existing building material (Cellulose insulation) to investigate the incidence of the water vapour permeability and the effect of the presence of a surface coating on the moisture buffering capacity.

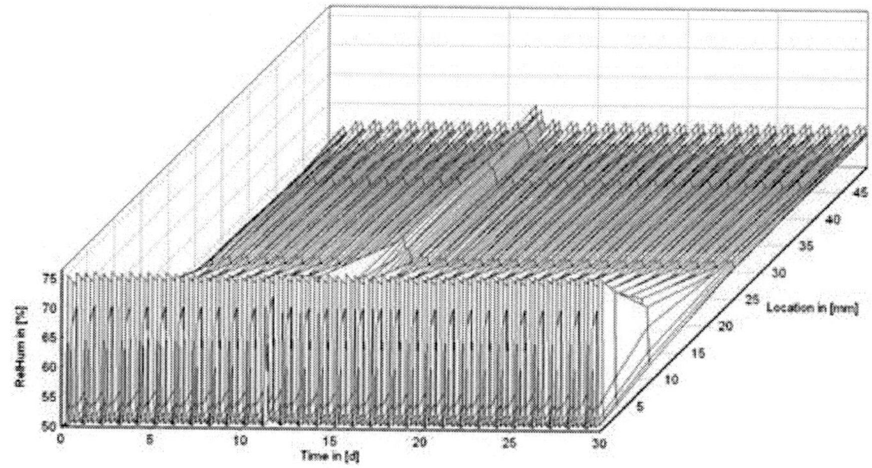

Figure 3. D visualization of the effect of the intermittent vapour sources on the material RH

Table2. Summary of simulation conditions

	Simulation	Thickness	Vapour resistance factor [-]	Surface exchange coefficient [s/m]
Case1: effect of thickness Boundary and test conditions : T=20°C Initial RH= 50% Simulation period= 30gg Daily vapour sources= 3	1	20	1,5*	$5{,}88\cdot10^{-8}$*
	2	30	1,5*	$5{,}88\cdot10^{-8}$*
	3	40	1,5*	$5{,}88\cdot10^{-8}$*
	4	50	1,5*	$5{,}88\cdot10^{-8}$*
	5	60	1,5*	$5{,}88\cdot10^{-8}$*
	6	70	1,5*	$5{,}88\cdot10^{-8}$*
	7	100	1,5*	$5{,}88\cdot10^{-8}$*
	8	120	1,5*	$5{,}88\cdot10^{-8}$*
	9	140	1,5*	$5{,}88\cdot10^{-8}$*
Case2 : effect of vapour permeability	10	50	1	$5{,}88\cdot10^{-8}$*
	11	50	3	$5{,}88\cdot10^{-8}$*
	12	50	4	$5{,}88\cdot10^{-8}$*
	13	50	5	$5{,}88\cdot10^{-8}$*
	14	50	10	$5{,}88\cdot10^{-8}$*
Case3 : effect of surface exchange coefficient	15	50	5	$5{,}88\cdot10^{-6}$
	16	50	5	$5{,}88\cdot10^{-7}$
	17	50	5	$5{,}88\cdot10^{-10}$
	18	50	10	$5{,}88\cdot10^{-6}$
	19	50	10	$5{,}88\cdot10^{-7}$
	20	50	10	$5{,}88\cdot10^{-10}$

* value for cellulose insulation

For this simulation the indoor RH levels are defined assigning initial conditions (RH=50%) and intermittent vapour sources that raise RH to 90% (Figure 3), while keeping the temperature constant (T=20°C).

3. RESULTS

3.1. Dynamic Characterization

Figure 4 summarizes the practical MBV of the tested materials. Sodium polyacrilate shows remarkable sorption capacities compared to the other tested materials. Cellulose, gypsum board and perlite follow in this order. The Moisture Buffer Values of both sodium polyacrilate (MBV= 9g/(m^2 %RH) @8/16h) and cellulose-based material (MBV= 3g/(m^2 %RH) @8/16h) are higher than the MBV of any other common building material tested in the NORDTEST Project [Rode, 2005] and they exhibit an "excellent" moisture buffer performance.

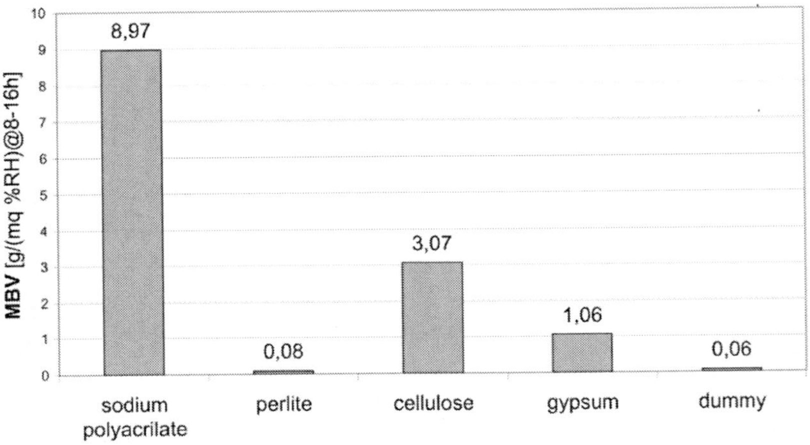

Figure 4. Practical MBV of the materials tested in the climatic chamber

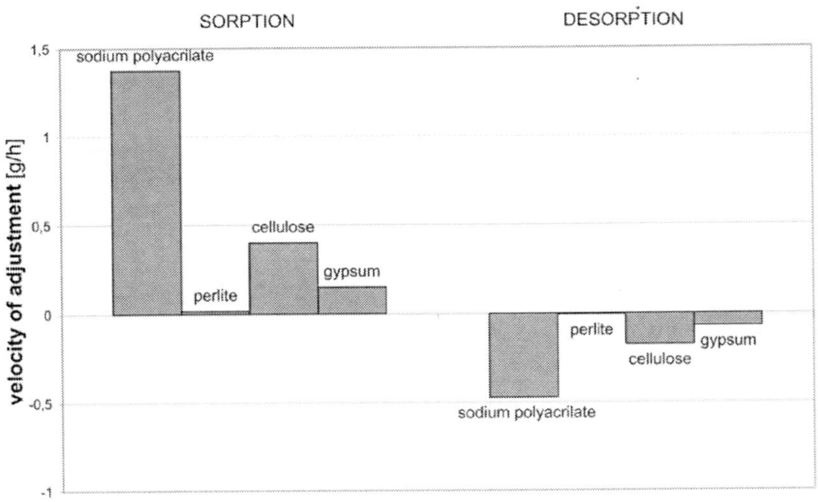

Figure 5. Velocity of adjustment of the materials tested in the climatic chamber to indoor air RH% variations (sorption and desorption phases)

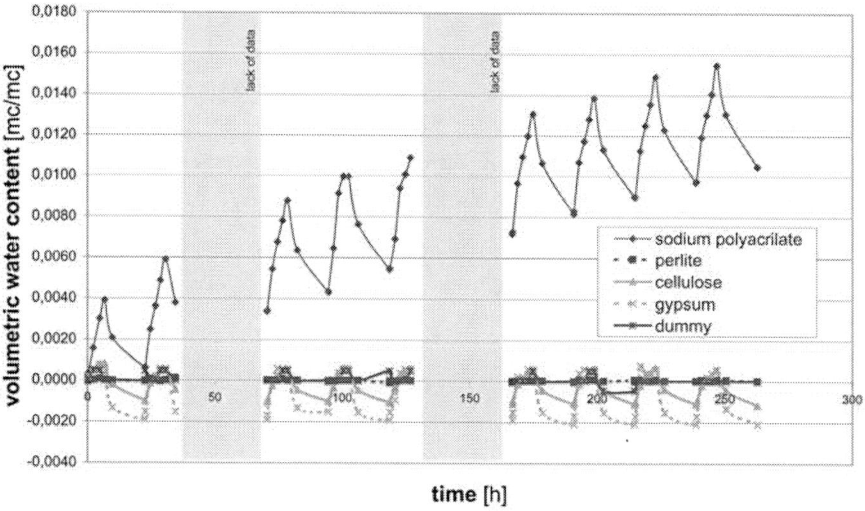

Figure 6. Volumetric water content of the materials during the periodical exposure of 8 h at 75% RH and 16 h at 33% in climatic chamber. For the sodium polyacrilate specimen, the minimum water content at the end of the desorption phase increases with the number of the cycles

The super absorbent polymer reacts more quickly than the other materials to RH% changes both in the adsorption and desorption phases. Cellulose-based material comes after (Figure 5).

The determination of the mass of moisture that remains in the specimen at the end of the desorption phase enables to evaluate the presence of hysteretic phenomena. In this sense the sodium polyacrilate is not able to discharge all the amount of water absorbed and therefore to maintain its sorption properties (Figure 6).

On the contrary the other tested materials don't show hysteresis.

The comparison of the amount of water stored up during the sorption phase and the maximum water content at equilibrium, expressed by the sorption isotherms, shows that (under experimental conditions), the materials are far from their functioning boundaries and may potentially meet with extreme RH% conditions or long term exposure.

3.2. Effect of Thickness and Material's Properties

Figure 7 shows the dependence of the maximum and final water content [kg_w/kg_s %] on the panel's thickness.

The maximum water quantity that is stored in the whole thickness in a certain moment is a measure of the sorption capacity of a material under simulation conditions. The maximum amount of water rises according to the panel thickening but the maximum water content, that is related to the weight of the material, decreases when the thickness increases from 20 to 50 mm and reaches a steady value above 50 mm.

On the other hand, the capacity of the material to retain the absorbed water at the end of the simulation period improves with the increase of thickness up to 50 mm.

Combining these acceptability ranges, we can notice that the variation of the material's thickness affects the maximum and final water content of the system till the thickness reaches 50mm. Beyond this value the increase of the material's thickness does not significantly contribute to the variation of the examined quantities and the system shows a steady response to thickness changes.

The final water content increases even with the vapour resistance factor μ [-] because the reduction of the water vapour permeability improves the capacity to retain water.

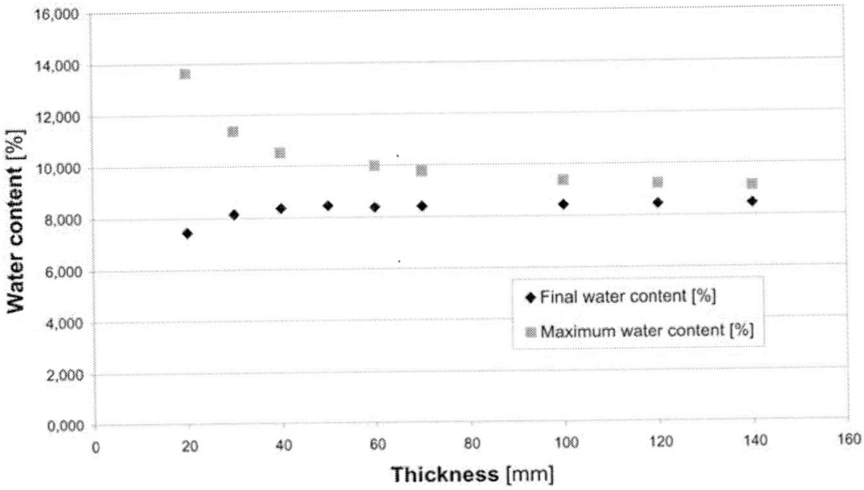

Figure 7. Simulated maximum and final water content (kg_w/kg_s%) stored up in a cellulose panel with variable thickness subject to intermittent vapour sources

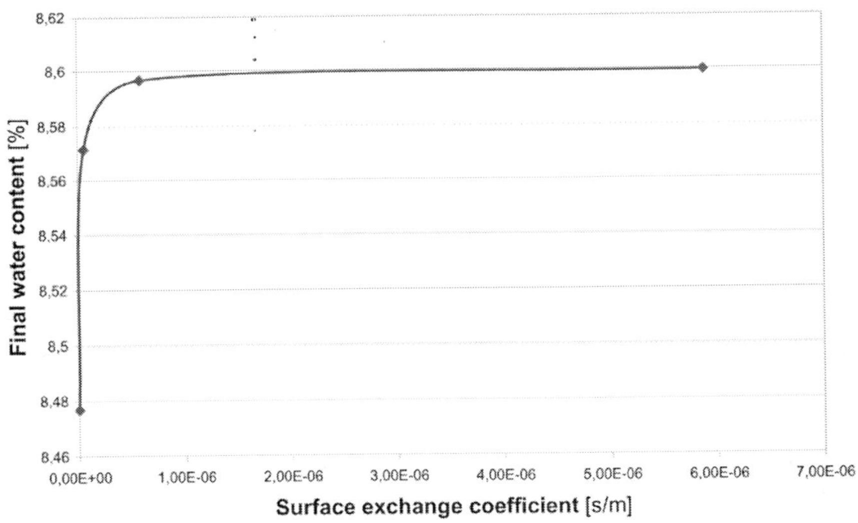

Figure 8. Simulated final water content (kg_w/kg_s%) for different surface exchange coefficients. Variations of 10, 100 and 10,000 times of Z_v of cellulose are investigated

Starting from these results, the surface exchange coefficient Z_v [s/m] is varied between $5,88\cdot10^{-6}$ s/m and $5,88\cdot10^{-10}$ s/m, emphasizing the influence of a different surface coating on the moisture buffering capacity of the hygroscopic material. As Figure 8 shows, the final water content slightly arises (about 0,1%) while Z_v increases 100 times from $5,88\cdot10^{-10}$ s/m to $5,88\cdot10^{-8}$ s/m and it is steady over this value. These data allow to conclude that the moisture buffering capacity is slightly influenced by the surface exchange coefficient and therefore if a surface coating or a packaging system is needed it will have no considerable effects on the material's sorption capacity.

Conclusion

It is possible to meet with the general reduction of the envelope's air permeability without the introduction of HVAC systems in occupied buildings. The introduction of small quantities of highly absorbing materials, as a part of the envelope components, furniture or furnishing, could dampen indoor RH% variations, improving comfort, air quality and healthiness.

Sodium polyacrilate and cellulose-based material exhibit an "excellent" moisture buffer performance and a high velocity of adjustment to RH% variations, that exceed the capacity of more common building materials. Nevertheless, sodium polyacrilate shows a hysteretic behaviour that in time may cause the loss of its sorption properties.

The cellulose-based material seems therefore to be the most suitable for moisture buffer applications.

Starting from the properties of a cellulose-based building material (*Cellulose insulation*) a numerical model is used to investigate the effect of thickness, vapour resistance factor and surface exchange coefficient on moisture buffer performance.

The analysis of the numerical results enables to define optimum functioning ranges for these quantities, that may guide the selection of a material with a potentially high-performance behaviour. On this point, the numerical data confirm that cellulose is a good moisture buffering material.

Acknowledgments

This work was developed by S. Cerolini and M. D'Orazio and is a part of a more general research coordinated by A. Stazi.

References

Fang, L; Clausen, G; Fanger, PO. Impact of temperature and humidity on the perception of indoor air quality, *Indoor Air*, 1998, 8, 80-90.

Grunewald, J. *Documentation of the numerical simulation program DIM3.1.* Theoretical fundamentals Institute of Building Climatology, Technical University of Dresden, Germany, 2000.

Hameury, S; Lundstrom, T; Contribution of indoor exposed massive wood to a good indoor climate: in situ measurement campaign, *Energy and Buildings*, 2004, 36, 281-292.

Hameury, S. Moisture buffering capacity of heavy timber structures directly exposed to an indoor climate: a numerical study, *Building and environment*, 2005, 40, 1400-1412.

Osanyintola, OF; Simonson, CJ. Moisture buffering capacity of hygroscopic building materials: Experimental facilities and energy impact, *Energy and Buildings*, 2006, 38 (10), 1270-1282.

Osanyintola, OF; Talukdar, P; Simonson, CJ. Effect of initial conditions, boundary conditions and thickness on the moisture buffering capacity of spruce plywood, *Energy and Buildings*, 2006, 38(10), 1283-1292.

Padfield, T. Humidity buffering of interior spaces by porous, absorbent insulation, *Danmarks Tekniske Universitet Kgs. Lyngby*, Electronic Publication, www.bkm.dtu.dk.

Padfield, T. The role of absorbent building materials in moderating changes of relative humidity, Ph.D thesis, *Department of Structural Engineering and Materials*, The Technical University of Denmark, November 1998, http:// www. natmus. dk/ cons/ tp/ phd/phd-indx.htm

Rode, C. *Moisture Buffering of Building Materials*, BYG DTU Report, Department of Civil Engineering, Technical University of Denmark, 2005.
http://www.byg.dtu.dk/upload/institutter/byg/publications/rapporter/byg-r126.pdf

Sedlbauer, K. Prediction of mould growth by hygrothermal calculation, *Journal of Thermal Envelope and Building Science*, 2002, 25, 321-335.

Sherman, MH; Matson, NE. *Air tightness of new U.S. houses: a preliminary report*, Environmental Energy Technologies Division, Energy Performance of Buildings Group, Indoor Environment Department, Lawrence Berkeley National Laboratory, Berkeley, 2002.

Simonson, CJ; Salonvaara, M; Ojanen, T. Improving indoor climate and comfort with wooden structures, *VTT Building Technology*, Espoo: VTT Publications 431, 2001, http://virtual.vtt.fi/inf/pdf/publications/2001/P431.pdf,VTT Building and Transport.

Simonson, CJ; Salonvaara, M; Ojanen, T. The effect of structures on indoor humidity – possibility to improve comfort and perceived air quality, *Indoor Air*, 2002, 12, 243-251.

Svennberg, K; Hedegaard, L; Rode, C. Moisture buffer performance of a fully furnished room, in: Proceedings (CD) of the Performance of Exterior Envelopes of Whole Buildings IX International Conference, *Clearwater Beach*, FL, 2004.

Svennberg, K. *Moisture Buffering in the Indoor Environment*, Doctoral Thesis, Report TVBH-1016 Lund 2006, Building Physics LTH.

Toftum, J; Jorgensen, AS; Fanger, PO. Upper limits of air humidity for preventing warm respiratory discomfort, *Energy and Buildings*, 1998, 28, 15-23.

Toftum, J; Jorgensen, AS; Fanger, PO. Upper limits for indoor air humidity to avoid uncomfortably humid skin, *Energy and Buildings*, 1998, 28, 1-13.

In: Relative Humidity: Sensors, Management... ISBN: 978-1-61761-734-8
Editor: John M. Castillo

Chapter 12

RELATIVE HUMIDITY MODELED BY A FUNCTIONAL NETWORK

Jose de Jesus Rubio[1*], Jaime Pacheco[1], Martin Salazar[2] and Raul Lugo[3]

[1]Seccion de Estudios de Posgrado e Investigacion- ESIME Azcapotzalco Instituto Politécnico Nacional. Av. De las Granjas, No. 682, Col. Sta. Catarina. Azcapotzalco, 02250 México D. F., México

[2]Division de Ingenieria Mecatronica Industrial. Tecnologico de Estudios Superiores de Ecatepec Av. Tecnologico, Col. Valle Anahuac, Ecatepec, Edo. de Mex

[3]Departamento de Ingenieria de Procesos e Hidraulica Universidad Autonoma Metropolitana - Iztapalapa. Av. Sn. Rafael Atlixco 186, Col. Vicentina Iztapalapa, Mexico D.F

ABSTRACT

The relative humidity will be modeled by a functional network. The relative humidity is based in real data obtained in a bird incubator. The functional networks compared with the neural networks have the advantage that they can find the global minimum while the neural networks can find a local minimum, but they have the disadvantage that they learn offline while some neural networks can learn online.

Keywords: Relative Humidity, Functional Networks.

1. INTRODUCTION

Some new neural networks are the functional networks, this can be considered as the elements which approximate a function, it cannot be updated using a common algorithm as

* Corresponding author: jrubioa@ipn.mx

the least square because it is complex. In this chapter the relative humidity in the range of 45% to 70% is approximated with a functional network.

2. Humidity

The humidity is that there exists vapor of water in the air.

2.1. Specific Humidity (ω or rh)

The specific humidity is a measure of the humidity in the air. It is defined as mass of the water vapor (m_v) contained in the mass of dry air (m_a).

$$\omega = \frac{m_v}{m_a} \tag{1}$$

2.2. Relative Humidity (ϕ)

The relative humidity of an air water mixture is defined as the ratio of the partial pressure of water vapor p_v in the mixture to the saturated vapor mixture of water p_s at a prescribed temperature. It is denoted with "z"

$$z = \frac{p_v}{p_s} 100\% \tag{2}$$

3. Comparison Between a Neural Network and a Functional Network

The Figure 1 shows the comparison between the neural network and the functional network. While the neural network needs the updating of the weights w_{ij} to decrease the error of the approximation, the functional network learns functions which make sure to find the global minimum of the approximation error [2]. In addition, while the activation functions for the first one are sigmoid or tanh, the activation functions for the second are base functions, this property makes sure that the base functions are linear independent which reduce the machine time.

The functional networks need to learn the behavior of the system, it can be made using two methods, the first one known as structural learning and the other known as parameters learning. The first one can be split to learn the initial behavior, it provides the first structure, the second is to simplify the network using functional equations, it provides a simplified structure; one example is the Figure 2 where in a) it is shown the initial scheme of a problem and in b) it is shown the simplified structure.

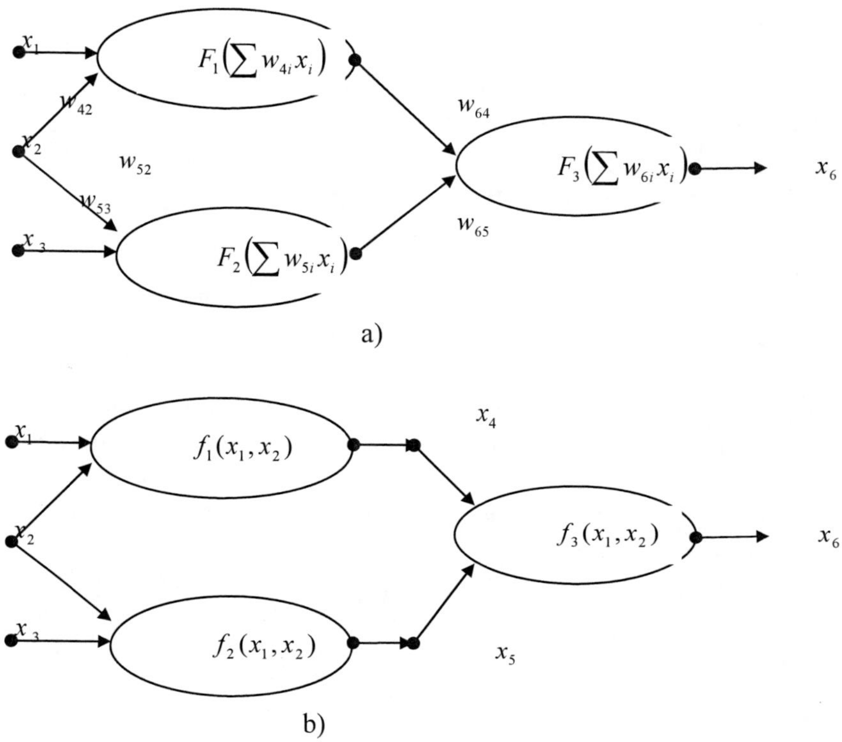

Figure 1. a) Neural network. b) Functional network

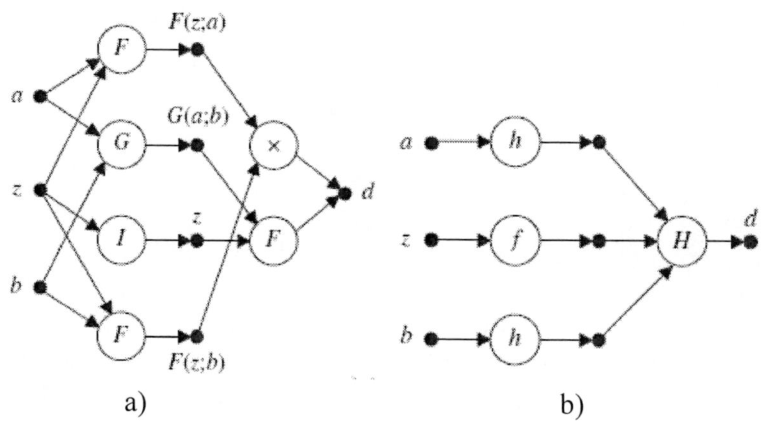

Figure 2. a) Initial funcional network. b) Simplified functional network

4. Elements of a Functional Network

The structure of a functional network is the connection between layers, the organization of the neurons and the properties of the final network [2], [5]. Please see the Figure 2 a).

One input layer of saved unities. It contains the input data (a, z, b).

One output layer of saved unities. It contains the output data (d).

One or many layers of processed unities. These unities are used to evaluate the input data from the previous layer and to provide the output data that will be used in the next layer. So, each neuron has a function which can be multi evaluated and can has as much arguments as inputs (F, G, I).

None or many hidden load layers. These are produced by functional unities, these layers let to force the coincidence of the output of the processing unities (F(z,a), G(b,a))

Set of managed links. They connect the input layer with the hidden layers (if they exist) and with the output layer, they show the processing flow of the information.

So, a *functional unity* on a set of nodes ***X*** is a tripled ***(Y, F, Z)*** where ***Y, Z*** are disjoints sets and not empty of ***X*** and ***F:*** $Y \rightarrow Z$ is a given function. The elements ***Y, Z*** and ***F*** are called the set of nodes of input, output and the processed function of the functional unity, respectively.

A *functional network* is a couple ***(X, U)***, where ***X*** is a set of nodes and ***U*** is defined as follows:

$$U = \{(Y_i, F_i, Z_i) \mid i = 1, \ldots, n\}$$

It is a set of functional unities on ***X*** which satisfy the following condition: each node $X_i \in X$ has to be and input or output node of at least one functional unity ***U***.

X can be split in 3 sets $\{X_{en}, X_{\text{int}}, X_{sal}\}$ where X_{int} can be an empty set, giving that the network is of one layer (one-layer).

5. Experimental Results

When the data of the functional network are known, it can be used to obtain the mathematical model of the relative humidity using the measures of the temperatures of the dry and the humidity bulbs. The Table 1 is used to obtain the mathematical model of the relative humidity using the measures of the temperatures of the dry and the humidity bulbs, in the more dark part of the Table there are the values where there are more born of birds.

From the Table 1 it is shown that the relative humidity changes between 47 % y 64 % for a temperature of the dry bulb of 36 °C a 38.8 °C, being it the optimal values for the born of birds. The function of the relative humidity are nonlinear, to obtain the mathematical model it is used the functional network known as the "Unity Model" [6]. Even this model is simple it can obtain many advantages because the parameters of this architecture are 3 functions where the output *z* can be described in function of the inputs *x, y* as follows:

$$z = f_3^{-1}(f_1(x) + f_2(y)) \tag{3}$$

It is not possible to simplify this model because there are not layer that are output of several functional unities. The training of the function $z = f_3^{-1}(f_1(x) + f_2(y))$ is equivalent to learn the functions $f_1(x)$, $f_2(x)$, $f_3(x)$. Then the learning is to find the functions using some data which are the triple $\{(x_{1j}, x_{2j}, x_{3j}) \mid x_{3j} = F(x_{1j}, x_{2j}); j = 1,\dots,n\}$. To make it is presented a linear method which can be obtained using the the following:

$$x_{3j} = f_3^{-1}(f_1(x_{1j}) + f_2(x_{2j})) \Leftrightarrow f_3(x_{3j}) = f_1(x_{1j}) + f_2(x_{2j}); j = 1,\dots,n$$

The error can be obtained using the following:

$$e_j = \tilde{f}_1(x_{1j}) + \tilde{f}_2(x_{2j}) - \tilde{f}_3(x_{3j}); j = 1,\dots,n$$

Table 1. Relative humidity of the dry and the humidity bulbs for an incubator of birds

Dry \Humid	29,4 °C	32,2 °C	35,0 °C	35,5 °C	36,1 °C	36,6 °C	37,2 °C	37,7 °C	38,3 °C	38,8 °C
20,0°C	41	31	23	22	21	19	18	17	16	15
20,5°C	44	34	25	24	23	21	20	19	18	17
21,1°C	47	36	28	26	25	23	22	21	20	18
21,6°C	50	39	30	28	27	25	23	22	21	20
22,2°C	53	41	32	30	29	27	26	25	23	22
22,7°C	56	44	34	33	31	30	28	27	25	24
23,3°C	60	47	37	35	33	32	30	29	27	26
23,8°C	63	50	39	37	36	34	32	31	29	28
24,4°C	66	53	42	40	39	36	34	33	31	30
25,0°C	70	55	44	42	40	38	37	35	33	32
25,5°C	73	58	47	45	43	41	39	37	36	34
26,1°C	77	62	49	47	45	43	41	39	38	36
26,6°C	80	65	52	50	49	46	44	42	40	38
27,2°C	84	68	55	52	50	48	46	44	42	40
27,7°C	88	71	57	55	53	51	49	46	45	43
28,3°C	92	74	60	58	55	53	51	49	47	45
28,8°C	96	78	63	61	59	56	54	51	49	47
29,4°C	100	81	66	64	61	59	56	54	52	50
30,0°C		85	69	67	64	61	59	57	54	52
30,5°C		89	72	70	67	64	62	59	57	55
31,1°C		92	76	73	70	67	65	62	60	57
31,6°C		96	79	76	73	70	67	65	62	60
32,2°C		100	82	79	76	73	70	68	65	63
32,7°C			86	82	79	76	73	71	68	65
33,3°C			89	86	83	79	76	74	71	68
33,8°C			93	89	86	83	80	77	74	71
34,4°C			96	93	89	86	83	80	77	74
35,0°C			100	96	93	89	86	83	80	77

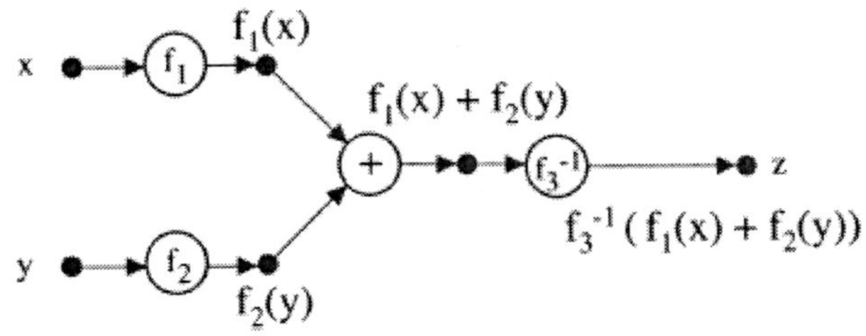

Figure 3. Unity model

So, each function of the network is obtained using a linear combination of known functions $\phi_s = \{\phi_{s1},\ldots,\phi_{sm}\}; s=1,2,3.$ and its initial conditions gives:

$$\widetilde{f}_s(x)=\sum_{i=1}^{m} a_{si}\phi_{si}(x); s=1,2,3.$$

$$\widetilde{f}_k(x_0)=\sum_{i=1}^{m_k} a_{ki}\phi_{ki}(x_0)=\alpha_K; k=1,2,3.$$

Where α_k are given constants. Obtaining the minimum of the errors, later using the Lagrange multipliers to obtain an auxiliary equation which is used to obtain the minimum [4].

$$Q=\sum_{j=1}^{n} e_j^2=\sum_{j=1}^{n}\left(\sum_{s=1}^{3}\sum_{i=1}^{m} a_{si}\phi_{si}(x_{sj})\right)^2$$

$$Q_\lambda=\sum_{j=1}^{n}\left(\sum_{s=1}^{3}\sum_{i=1}^{m} a_{si}\phi_{si}(x_{sj})\right)^2+\sum_{k=1}^{3}\lambda_k\left(\sum_{i=1}^{m_k} a_{ki}\phi_{ki}(x_0)-\alpha_k\right)$$

$$\frac{\partial Q_\lambda}{\partial a_{tr}}=2\sum_{j=1}^{n}\left(\sum_{s=1}^{3}\sum_{i=1}^{m} a_{si}\phi_{si}(x_{sj})\right)\phi_{tr}(x_{jt})+\lambda_t\phi_{tr}(x_0)=0$$

$$\frac{\partial Q_\lambda}{\partial \lambda_t}=\sum_{i=1}^{m_t} a_{ti}\phi_{ti}(x_0)-\alpha_t$$

$$t=1,2,3; \qquad r=1,\ldots,m$$

The mentioned above system has *3m+1* equations with *3m+1* unknown variables. It is assumed that the set of approximate functions $\{\phi_{ij}; i=1,\ldots,m\}; j=1,2,3.$ are linearly independent, the matrix of the system is nonsingular and it has unique solution. The approximation of the functions for the unity model using the prediction of the relative humidity is:

$$\left.\begin{aligned} f_1(x)&=0.950306+0.0496945x;\\ f_2(x)&=1.04978-0.047837x;\\ f_3(x)&=0.989221+0.0107788x;\end{aligned}\right\} \qquad (4)$$

It can be seen that the learning of the functions was made using a family of polynomials of the first degree because this approximation gives a small error and it uses less computation in the operations. Considering the relative humidity and the dry bulb the structure of the functional network is as follows:

$$\left.\begin{aligned} z &= f_3^{-1}(f_1(x)+f_2(y)) \\ z &= \frac{f_1(x)+f_2(x)-0.989221}{0.0107788} \\ z &= 92.7746(1.01087+0.0496945x_1-0.0497837x_2) \end{aligned}\right\} \quad (5)$$

Where:
$x_1 \rightarrow$ Temperature of the humidity bulb.
$x_2 \rightarrow$ Temperature of the dry bulb.
$z \rightarrow$ Relative humidity.

As an example of the approximation of the polynomial functions, let us consider that the temperature of the humidity bulb is 24.4°C and the temperature of the dry bulb is 37.2°C, substituting the values in the equation (5) gives:

$$\begin{aligned} z &= 92.7746(1.01087+0.0496945x_1-0.0497837x_2) \\ z &= 92.7746(1.01087+0.0496945(24.4)-0.0497837(37.2)) \\ z &= 34.46225331\% \approx 34\% \end{aligned}$$

Now, let us consider a temperature of the dry bulb of 38.8°C and a temperature of the humidity bulb of 30°C, substituting the values in the equation (5) gives:

$$\begin{aligned} z &= 92.7746(1.01087+0.0496945x_1-0.0497837x_2) \\ z &= 92.7746(1.01087+0.0496945(30)-0.0497837(38.8)) \\ z &= 52.89056196\% \approx 53\% \end{aligned}$$

Comparing the obtained results with the obtained in the Table 1, gives that the error is small and the functional network of the equation (5) is good to approximate the relative humidity.

CONCLUSIONS

The relative humidity was modeled by a functional network. The relative humidity is based in real data obtained in a bird incubator. The functional networks compared with the neural networks have the advantage that they can find the global minimum while the neural networks can find a local minimum, but they have the disadvantage that they learn offline while some neural networks can learn online.

REFERENCES

[1] Castillo, E., Cobo A., Gómez, R. & Hadi, A. (1999). functional equations; *neural networks*, Net-works *Vol. 35* No. 1, 70 - 82.

[2] Castillo, E., Cobo, A., Gutierrez, M. J. & Pruneda, E. (2000). Functional networks: A new network-based methodology, *Computer-Aided Civil and Infrastructure Engineering*, *Vol. 15*, no. 2, 90-106.

[3] Castillo, E., Gutierrez, J. M., Hadi, A S. & Lacruz, B. (2001). *Some Applications of Functional Networks in Statistics and Engineering, Technometrics*, *Vol. 43*, No. 1, 10-24.

[4] Kirk, D. E. (1970). Optimal Control Theory: An introduction. *Englewood Cliffs*, New Jersey: Prentice-Hall.

[5] Lacruz, B. & Perez, A. (2006). "Functional networks for classification and regression problems", *International Conference on Mathematical and Statistical Modeling.*

[6] Yong-Quan, Z. & Li-Cheng, J. (2006). One-variable Interpolation Function Based on Functional Networks, *International Journal of Information Technology*, *Vol. 12* , No. 2, 120-129.

INDEX

A

B

C

D

G

H

I

J

K

L

M

N

O

P

Q

R

S

T

U

V

W

X

Y

Z